Biostatistics for Bioassay

In recent decades, there has been enormous growth in biologics research and development, with the accompanying development of biological assays for emerging products. In parallel, there have been substantial advances in statistical methodology, as well as technological advances in computer power, enabling new techniques to be implemented via statistical software. ***Biostatistics for Bioassay*** presents an overview of the statistical analysis techniques that are needed in order to report the results of biological assays. These assays are needed for testing all biological medicines, such as vaccines and cell therapies, to allow them to be released for use. Beginning with consideration of the performance characteristics required of a bioassay, including accuracy, precision, and combinations of these two attributes, the book builds a framework for statistical bioassay design.

Features:
- Explains the statistical methods needed at each stage of the lifecycle of a bioassay
- Describes the demonstration of the bioassay's performance, known as validation
- Covers the statistical techniques for monitoring the bioassay's performance over time
- Details how to transfer the bioassay to another laboratory or replace critical reagents
- Provides examples at every stage, to allow the reader to work through the techniques and consolidate their understanding

The book provides a resource for interested bioassay analysts, and statisticians working with bioassays. In bringing together best practices in statistics across the bioassay lifecycle into a single volume, it aims to provide a comprehensive and useful textbook for statistical analysis in bioassay.

Ann Yellowlees is the Founder of Quantics Biostatistics. Ann has worked in academia, Shell Research, and the NHS in Scotland. She has been consulting in bioassay statistics since 2006. She holds degrees in both mathematics and applied statistics from the University of Oxford, and a PhD in statistics from the University of Waterloo.

Matthew Stephenson is a Director of Statistics at Quantics Biostatistics, with an undergraduate degree in human kinetics, and an MSc and PhD in statistics from the University of Guelph. He won the 2020 Canadian Journal of Statistics Award for his PhD research on leveraging graphical structure among predictors to improve outcome prediction.

Chapman & Hall/CRC Biostatistics Series
Series Editors: Mark Chang, *Boston University, USA*

Quantitative Methods for Precision Medicine
Pharmacogenomics in Action
Rongling Wu

Drug Development for Rare Diseases
Edited by Bo Yang, Yang Song and Yijie Zhou

Case Studies in Bayesian Methods for Biopharmaceutical CMC
Edited by Paul Faya and Tony Pourmohamad

Statistical Analytics for Health Data Science with SAS and R
Jeffrey Wilson, Ding-Geng Chen and Karl E. Peace

Design and Analysis of Pragmatic Trials
Song Zhang, Chul Ahn and Hong Zhu

ROC Analysis for Classification and Prediction in Practice
Christos Nakas, Leonidas Bantis and Constantine Gatsonis

Controlled Epidemiological Studies
Marie Reilly

Statistical Methods in Health Disparity Research
J. Sunil Rao

Case Studies in Innovative Clinical Trials
Edited by Binbing Yu and Kristine Broglio

Value of Information for Healthcare Decision Making
Edited by Anna Heath, Natalia Kunst, and Christopher Jackson

Probability Modeling and Statistical Inference in Cancer Screening
Dongfeng Wu

Development of Gene Therapies
Strategic, Scientific, Regulatory, and Access Considerations
Edited by Avery McIntosh and Oleksandr Sverdlov

Bayesian Precision Medicine
Peter F. Thall

Statistical Methods for Dynamic Disease Screening and Spatio-Temporal Disease Surveillance
Peihua Qiu

Causal Inference in Pharmaceutical Statistics
Yixin Fang

Applied Microbiome Statistics
Correlation, Association, Interaction and Composition
Yinglin Xia and Jun Sun

Association Models in Epidemiology
Study Design, Modeling Strategies, and Analytic Methods
Hongjie Liu

Likelihood Methods in Survival Analysis
with R Examples
Jun Ma, Malcolm Hudson, and Annabel Webb

Biostatistics for Bioassay
Ann Yellowlees and Matthew Stephenson

For more information about this series, please visit: www.routledge.com/Chapman--Hall-CRC-Biostatistics-Series/book-series/CHBIOSTATIS

Biostatistics for Bioassay

Ann Yellowlees and Matthew Stephenson

CRC Press
Taylor & Francis Group
Boca Raton London New York

CRC Press is an imprint of the
Taylor & Francis Group, an **informa** business

A CHAPMAN & HALL BOOK

First edition published 2025
by CRC Press
2385 Executive Center Drive, Suite 320, Boca Raton, FL 33431, U.S.A.

and by CRC Press
4 Park Square, Milton Park, Abingdon, Oxon, OX14 4RN

CRC Press is an imprint of Taylor & Francis Group, LLC

ISBN: 978-1-032-57972-6 (hbk)
ISBN: 978-1-032-58242-9 (pbk)
ISBN: 978-1-003-44919-5 (ebk)

DOI: 10.1201/9781003449195

Typeset in CMR10 font
by KnowledgeWorks Global Ltd.

Publisher's note: This book has been prepared from camera-ready copy provided by the authors.

For our families, and with our thanks to long-suffering colleagues at Quantics.

Contents

III   Validation, monitoring and modifications   177

8   Qualification and validation of bioassays   179

9   Monitoring the performance of a bioassay procedure   213

Preface

Statistical Method in Biological Assay [19], by David Finney, was first published in 1952 and subsequently revised in 1964 and 1978. The book has been a go-to text, particularly for statisticians working in bioassay. Finney's aim was *'to develop the theory and methods as far as seems practicable at this time'* [19].

Since the last edition of Finney's book, there have been major advances in three relevant areas. First, there has been enormous growth in biologics research and development, with accompanying development of biological assays for the emerging products. Thus, widespread competence in the statistical analysis of biological assays, or bioassays, is needed. There are not, at this time, enough professional statisticians to handle this growth in demand. Second, there have been substantial advances in statistical methodologies; and third, technological advances in computer power and statistical software have enabled new statistical techniques to be implemented.

Statistical guidance documents for the analysis of bioassay data have been produced by regulatory bodies, each providing information about best practice on specific topics across the design, development and analysis of bioassays. However, as far as we are aware, there is no single textbook that brings all aspects together in one place.

Our aim in this book is twofold. We want to provide a single source of information about bioassay statistics for trained statisticians who are new to bioassay data analysis. We also want to explain bioassay statistical analysis at a level of detail that is useful for bioassay scientists with an interest in understanding the analysis provided for them by a statistician, or by a statistical package, or those interested in conducting the analysis themselves.

We present and explain the statistical methods needed at each stage of the lifecycle of a bioassay. In Chapter 1, we provide an introduction to the concept of a bioassay and the reasons for the need for special statistical treatment of the evaluation of a biological product. The use of a reference standard in bioassays is introduced and the key concept of a dose-response relationship, which underlies bioassay design, is outlined. The mathematically simple concentration-determining assay, where the concentration of an unknown test sample is determined by reading off from a standard curve, is described. The more complex case, where both the test sample and the reference standard are evaluated over a range of doses, is considered, and the concepts of relative potency and absolute potency are introduced. Relative potency is defined, following Finney, in terms of the dose-response relationships of the reference standard and the test sample, as the ratio of doses that result in the same response. The concept of parallel line or parallel curve dose-response relationships is introduced and its fundamental importance to relative potency is explained. Effective dose, ED_{50}, and effective concentration, EC_{50}, are defined.

Chapter 2 provides a discussion of the purpose of a bioassay and the requirements placed on the bioassay's performance. We provide a description of the performance characteristics that are measured to determine whether the bioassay is fit for a particular purpose – for example, batch release. The analytical target profile is introduced. Performance characteristics, including accuracy and precision, as well as linearity, range and combinations of accuracy and precision, are defined. The statistical properties of these characteristics are explained. We explore the combined impacts of manufacturing variability and measurement variability on the bioassay-reported result.

Chapter 3 covers the design and optimisation of the bioassay with a view to achieving the required performance in a given application. The focus of the chapter is on randomisation and replication of samples, both within a bioassay session and between sessions.

In Chapter 4, we introduce the mathematical models typically used to describe dose-response relationships in bioassays. We cover both continuous (quantitative) and binary (qualitative) response types, and we demonstrate the calculation of relative potency for each model. Chapter 5 considers the choice of dose-response relationship in a given situation. The statistical assumptions that are required to allow inferences to be made about the potency of a sample are defined. Methods are presented for determining whether or not the assumptions are justified. In the case where they are not justified, solutions are provided. The number of doses to be used, and their spacing, are discussed.

The statistical analysis of an individual bioassay is covered in Chapter 6. This includes fitting the selected dose-response relationship to the data in a range of scenarios, as well as detecting potential extreme data points or outliers. As well as the most commonly used statistical models for continuous data, we also include the use of mixed models to account for structure in the bioassay. The logit model for binary data is discussed, and we mention the analysis of time-to-event data.

The validity of an individual bioassay measurement, via system and sample suitability criteria, is examined in Chapter 7. We begin by outlining the two standard approaches to assessing suitability: significance testing and equivalence testing. We discuss methods for setting equivalence bounds for the criteria in a range of scenarios, including the assessment of the fit of the dose-response relationship and the parallelism of the reference standard and test samples.

Chapter 8 covers the assessment of the performance of a bioassay, or bioassay qualification and validation. We discuss the choice of acceptance criteria for the bioassay performance in line with the analytical target profile. The design of qualification and validation studies follows, including the choice of test samples and the number of replicate runs required to provide the required probability of a successful study. Two fully worked examples are included.

In Chapter 9, we introduce monitoring the performance of a bioassay procedure over its lifetime. Techniques of statistical process control are described, and appropriate metrics and monitoring rules are discussed. We consider actions that may be required following a breach of rules, for example the detection of a trend in a parameter. Finally, in Chapter 10, we consider changes in the bioassay that are likely to be encountered over the course of its lifetime. Such changes can affect the performance of the bioassay and they need to be introduced with care. They include the transfer of a bioassay method from one laboratory to another and introduction of new reference standards and other critical reagents. We discuss the design of studies to ensure appropriate transition in these scenarios.

Appendix A covers the basic statistical methodologies required, and Appendix B provides the data sets used in the examples so that readers can reconstruct the analyses presented in the book.

The majority of the analyses described can be implemented in Excel [38]. Exceptions include iterative procedures for fitting non-linear dose-response models to bioassay data: for these, statistical software or bioassay-specific software is required. In addition to Excel, in the preparation of this book we used R software [49]. The `ggplot2` package [68] was used to produce all plots; the `VCA` package [54] was used to conduct variance components analysis; the `lme4` package [2] was used to fit mixed models; and the `minpack.lm` package [16] was used to fit non-linear dose-response models. We also made use of QuBAS bioassay software [48] for the analysis of dose-response data.

It is important to note that, in the examples, no rounding was done until the final result is presented. Where intermediate results are provided, these are rounded for presentation purposes. Therefore, a reconstruction of any final result can only be expected to match the result presented if the original raw data, as provided in Appendix B, are used. Where results from previous analyses are used in examples, the rounded values have been used as noted in the text.

Authors

Ann Yellowlees

Ann Yellowlees is a founder of Quantics Biostatistics, and a USP Expert Committee Member (Statistics). She holds an MA in mathematics and an MSc in applied statistics, both from the University of Oxford, and a PhD in statistics from the University of Waterloo. She is a fellow of the Royal Statistical Society and a Chartered Statistician (CStat). A spell in academia at the University of Bristol followed completion of her PhD, after which she joined Shell Research, UK, where she headed the Mathematics and Statistics section. She then joined the Information and Statistics Division of the National Health Service in Scotland to lead the Cancer Surveillance team. She established Quantics, a statistical consultancy specialising in life sciences, in 2002, and has specialised in bioassay statistics since 2006. Drawing on experience in other fields of statistics and mathematics, she has authored several peer-reviewed publications in biostatistics for bioassay.

Matthew Stephenson

Matthew Stephenson is the Director of Statistics at Quantics Biostatistics, where he has been advising clients on the statistical analysis of bioassays since 2019. He holds an undergraduate degree in human kinetics and an MSc and PhD in statistics from the University of Guelph. He was awarded the 2020 Canadian Journal of Statistics Award for his PhD research on leveraging the graphical structure among predictors to improve outcome prediction. Following a stint as Assistant Professor in Statistics at the University of New Brunswick, he resumed a full-time role at Quantics in 2023.

Abbreviations

Abbreviation	Meaning
4PL	Four-parameter logistic
5PL	Five-parameter logistic
ANOVA	Analysis of variance
ATP	Analytical target profile
CFD	Critical fold difference
CI	Confidence interval
Cpm	Process capability index
$C_S^{(N)}$	Concentration of new reference standard
$C_S^{(O)}$	Concentration of original reference standard
C_T	Concentration of test sample
CUSUM	Cumulative sum
CV	Coefficient of variation
EC_{50}	Effective concentration at 50% response
ED_{50}	Effective dose at 50% response
EWMA	Exponentially weighted moving average
GCV	Geometric coefficient of variation
GMP	Good manufacturing practice
I-chart	Individual values chart
IP	Intermediate precision
LCL	Lower confidence limit
LLOQ	Lower limit of quantitation
LOD	Limit of detection
LSL	Lower specification limit
MR-chart	Moving range chart
MSLoF	Mean squared error: lack-of-fit
MSPE	Mean squared error: pure error
MSS	Mean sum of squares
OOS	Out of specification
PF	Precision factor
Prob(OOS)	Probability of an out of specification result
QC	Quality control
QQ	Quantile-quantile
RA	Relative accuracy
RB	Relative bias
R-chart	Range chart

RMSE   Root mean squared error
RP   Relative potency
RSS   Residual sum of squares
RU   Receiving unit
RV   Reportable value
S   Reference standard
S-chart   Standard deviation chart
SD   Standard deviation
SSE   Sum of squared errors
SSLoF   Sum of squared errors: lack-of-fit
SSPE   Sum of squared errors: pure error
SU   Sending unit
T   Test sample
TAE   Total analytical error
TAP   Transfer of analytical procedure
TI   Tolerance interval
TRP   True relative potency
UCL   Upper confidence limit
ULOQ   Upper limit of quantitation
USL   Upper specification limit
USP   United States Pharmacopeia

Notation

Notation	Meaning
A	Left asymptote of the 4 and 5 parameter logistic distributions
B	Slope parameter of the 4 and 5 parameter logistic distributions
C	Location parameter of the 4 and 5 parameter logistic distributions
D	Right asymptote of the 4 and 5 parameter logistic distributions
E	Asymmetry parameter of the 5 parameter logistic distribution
G	Grubbs' test statistic
H_A	Alternative hypothesis
H_0	Null hypothesis
k_F	Number of runs in a session: procedure format
k_s	Number of runs in a session: validation study
n_F	Number of sessions: procedure format
n_s	Number of sessions: validation study
$\text{Off}^{(N:O)}$	Offset: new reference standard versus original reference
P	Proportion of the reportable value variance due to process variance
R	Response value
RB^+, RB^-	Pair of equivalent positive and negative RB values
$RP^{(N)}$	Relative potency with respect to new reference standard
$RP^{(O)}$	Relative potency with respect to original reference standard
α	Type I error rate
β	Type II error rate
β_0	Intercept of a linear relationship
β_1	Slope of a linear relationship
β_2	Quadratic term of a quadratic relationship
ϵ	Random error term
$\mu_{\log RP}$	Mean value of measured log RP for a given sample
μ_{manuf}	Mean value of measured log RP for the manufacturing distribution
μ_{product}	Mean value of true log RP for the product manufacturing process
$\Phi(\cdot)$	Cumulative normal distribution function
$\sigma^2_{\log RP}$	Variance of measured log RP for a given sample
$\sigma^2_{\text{Between sessions}}$	Between session variance
σ^2_{manuf}	Variance of measured log RP for the manufacturing distribution
$\sigma^2_{\text{product}}$	Variance of measured log RP for the product manufacturing process
$\sigma^2_{\text{Within session}}$	Within session variance

NB: 'Hat' notation is used to indicate that a quantity is an estimate based on a data set. For example, $\hat{\sigma}^2_{\text{Between sessions}}$ means the estimate of $\sigma^2_{\text{Between sessions}}$.

Part I

Potency and its measurement: bioassays and their performance

1

Introduction – bioassays and potency

Unlike small-molecule drugs, which are relatively simple chemical structures that can be synthesised in the laboratory, a biological medicine (or 'biologic') is structurally very complex and is produced in living cells from which it is then isolated.

The quality and potency of a new batch of a small-molecule drug can be determined by ensuring the synthesis process creates the correct chemical structure and examining the purity of the final product. In contrast, because of the structural complexity and method of production, the potency of a new batch of a biologic must usually be determined by examining its biological activity rather than its physio-chemical properties. Such tests are called biological assays or bioassays. Typically, bioassays assess the biological activity of a biologic drug or batch of a drug by comparison with the known activity of a reference standard [58].

The first use of a bioassay dates back to the late 19th century, when the foundation of bioassays was laid down by German physician Paul Ehrlich [64]. Many early bioassays involved animal studies, but the majority now use organ, tissue or cell culture systems, or combinations of these. These assays seek to reflect the biological mechanism of action of the drug so that a connection can be made between the assay result and its therapeutic effect.

Most countries will only accept the import and sale of medicines that have been manufactured to internationally recognised good manufacturing practice (GMP) – a system for ensuring that products are consistently produced and controlled according to quality standards [47]. Every batch of biological medicine produced by a manufacturer must undergo rigorous and independent testing before it can be released onto the market for human use [22]. For regulatory approval of biological products such as vaccines, monoclonal antibodies, and cell and gene therapies, a validated bioassay for evaluating the potency of the product or the concentration of an analyte must have been accepted by the regulator. This bioassay will then be used at batch release and for the monitoring of the product's stability. It may also be used to support changes in the production process.

Because both the manufacturing process for a biological product and the bioassay system are variable, there is generally considerable variability among measurements obtained by apparently identical operations. Statistical methods are needed to capture and characterise the variability and ensure that the maximum information is extracted from the available data. This is important from the point of view of cost but also for minimising the use of live animals until alternatives can be found.

DOI: 10.1201/9781003449195-1

1.1   Structure of a bioassay

Bioassays, in general, seek to understand the relationship between a dose of a product and the response it elicits. Finney [19] states:

> *The typical bioassay involves a stimulus (for example, a vitamin, a drug, a fungicide) applied to a subject (for example, an animal, a piece of animal tissue, a plant, a bacterial culture). The intensity of the stimulus can be varied ... so as to vary the dose given to the subject, and this dose can be measured. Application of the stimulus is followed by a change in some measurable characteristic of the subject, the magnitude of the change being dependent on the dose. A measurement of this characteristic is the response of the subject.*

Various responses can be measured in a bioassay and will depend on the product. Interest often lies in understanding the properties of a new batch of a drug; this is commonly referred to as a test sample. For example, the efficacy of a new batch of a vaccine could be measured by exposing vaccinated animals to the virus and measuring the proportion of animals in each dose group that were disease-free at the end of the study. This type of study is known as a challenge assay. In a cell-based assay, cellular products specific to the mechanism of action may be tagged so that luminescence can be measured.

Figure 1.1 provides example plots of possible relationships between dose and response. Note that it is most common for the dose to be plotted on a logarithmic rather than a linear scale. That is, the plotted doses increase in a multiplicative (rather than additive) fashion. To provide an example, when plotted on a logarithmic scale, the distance between doses of 2 and 4 units is the same as the distance between doses of 4 and 8 units (Figure 1.1(a)); on a linear scale the distance between doses of 2 and 4 units is half the distance between doses of 4 and 8 units. The nature of the relationship will often exhibit an 'S'-shaped curve with lower and upper limiting responses. For example, in the case of a challenge assay, the upper limiting response occurs when all animals are disease-free (100%), and the lower limiting response corresponds to none of the animals being disease-free (0%). For a cell-based assay measuring luminescence, these limits might be characteristics of the measuring device.

Most often the 'S'-shaped dose-response curve will be symmetric (Figure 1.1(a)), but for some bioassays the relationship will be asymmetric with the curve being steeper at either the top or bottom (Figure 1.1(b)). In practice, it may not be possible to observe the entire dose-response relationship and instead only the linear portion of the curve may be realised in the bioassay (Figure 1.1(c)). When the response variable is binary, then the proportion of observations falling in a particular category will be modelled (Figure 1.1(d)).

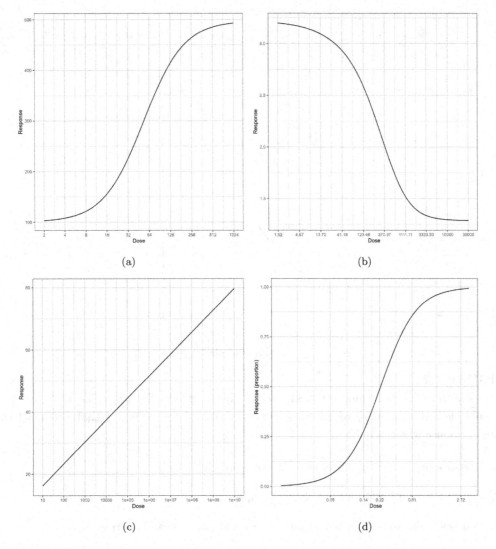

FIGURE 1.1

Plots of dose-response models: a symmetrical 'S'-shaped curve in (a); an asymmetrical 'S'-shaped curve in (b); a straight line in (c); and a binary response in (d). For all these relationships, the dose is plotted on a logarithmic scale.

1.2   Result of a bioassay

Bioassays are used to understand various characteristics of biologic compounds. The choice of result will be context-specific and depend on the purpose of the bioassay. We will discuss three different types of bioassay:

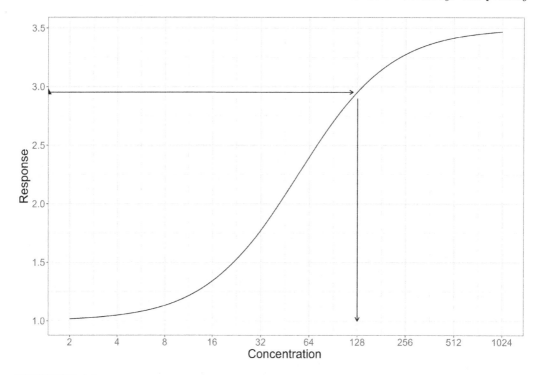

FIGURE 1.2

Interpolation of a test sample concentration from a standard curve. The unknown test sample yielded a response of 2.95; the concentration of the standard that produced the same response was 128 units.

1. Presence or absence of material in the test sample;

2. Concentration of material in the test sample;

3. Potency – the amount of test sample needed to produce a given response.

1.2.1   Presence or absence of material

The simplest result for a bioassay is non-quantitative and simply seeks to report the presence or absence of a particular material or substance in the test sample. This could be, for example, the detection of impurities resulting from the manufacturing process and may be based on a quantitative result. If the bioassay has a limit of detection, then interest may be in whether or not the value is above or below this limit.

1.2.2   Concentration of material

To assess the concentration of material in a sample, whether impurity or active ingredient, a known standard sample can be used as a reference. This known standard may, for example, be provided by a manufacturer of an assay kit. The concentration of the standard sample that produces the same response as the test sample is taken to be the estimate of the unknown concentration. For example, in Figure 1.2, the concentration of the standard that

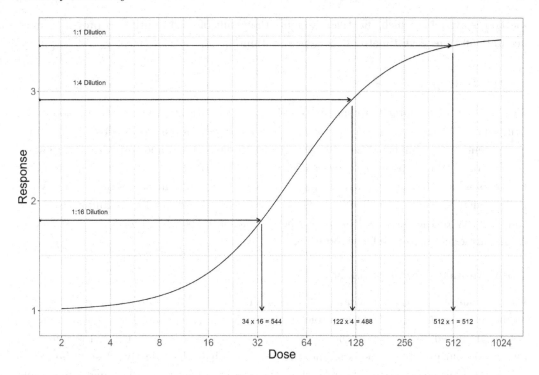

FIGURE 1.3

Interpolation of a test sample concentration at full, $1/4$ and $1/16$ strength from a standard curve. The estimated concentration for the 1:1 dilution was 512; 488 for the 1:4 dilution; and 544 for the 1:16 dilution. Combining these three estimates using the geometric mean gives a concentration estimate of 514.2.

produces the same response as the unknown test sample, and therefore the estimate of the unknown concentration, is 128 units. Bioassays such as this are known as bioanalytical or interpolation assays.

In an interpolation bioassay, the test sample can be tested one or more times at full strength. Additionally, it can be diluted, for example to half-strength and to quarter-strength. These diluted samples can also be tested in the bioassay, and the results can be combined to provide the final result. Figure 1.3 shows an example where the unknown test sample was tested once at full strength and again at $1/4$ and $1/16$ strength. A simple (conventional) approach to combining the concentrations from different test sample dilutions is to multiply each concentration by the relevant dilution factor (4 and 16, respectively), then calculate the geometric mean to produce the final concentration estimate. For the example in Figure 1.3, the combined estimate of the concentration is 514.2 units.

Dilution can be useful when results fall close to the upper limit of the standard sample, making the readout unreliable. A similar approach can be taken if the test sample responses lie towards the lower end of the standard curve by concentrating the test sample.

The conventional method of multiplying by the dilution factor and then combining the results across dilutions does not account for the reliability of each estimate. An alternative method is available which effectively weights the result of each dilution to provide a more accurate and precise estimate of the concentration [9].

1.2.3   Potency

The potency of a material refers to its ability to produce a given effect or response. A drug that is highly potent may produce a particular response at a low dose, while, for a less potent drug, a higher dose will be required to yield that same response. There are a number of ways to express potency, and characterising the potency is critical in the batch release of a biologic drug. We will introduce the idea of absolute measures of potency, where the potency is assessed on the observed response to the (unknown) test sample. However, the majority of the book will focus on the estimation of potency relative to a standard, commonly referred to as relative potency.

Absolute measures of potency

Potency can be expressed in a range of ways with respect to the dose-response relationship of the test sample alone. Two examples include: the EC_{50} which is the 'effective concentration' required to produce a biological response that is halfway between the minimum and maximum possible responses; and the ED_{50} which is the 'effective dose' required to produce a biological response in 50% of the population of interest. These measures apply to a single sample without reference to another sample, and we refer to them as absolute potency measures.

Figure 1.4 illustrates the potency measures EC_{50} and ED_{50}. Although these graphs appear similar, the underlying data structures are different and the analysis of the data needs to be handled differently. In Figure 1.4(a), the response is a continuous variable and so can take on any value between the asymptotes. In Figure 1.4(b), the response for each subject is binary. Therefore, the response for a dose group of, for example, size 20 can only take on values of 0, $1/20 = 0.05$, $2/20 = 0.1$, and so on up to $20/20 = 1$.

1.3   Relative potency

Bioassay systems are highly variable, and the estimated concentration or potency of a test sample can be a very variable quantity from run to run of the bioassay. There are many factors that may contribute to the variability of the bioassay result. These factors may be known, such as incubation time or incubation temperature, but can also include other unknown factors. While some factors might be easy to control, other factors may be near impossible to account for. Some of this variability can be captured by including a reference sample (with known concentration) in every run of the bioassay. Many of the factors that cause the variability of a biological assay from run to run will affect the reference and test samples in similar ways. We can expect samples to change similarly to each other within a run. That is, their dose-response curves are likely to move in tandem. Figure 1.5 illustrates

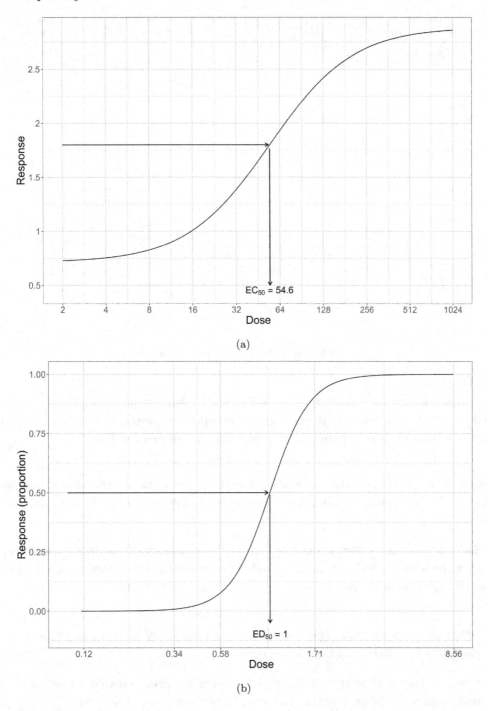

FIGURE 1.4

The EC_{50} is the effective concentration required to produce a biological response that is halfway between the minimum and maximum possible responses in (a). The ED_{50} is the effective dose required to produce a biological response in 50% of the population of interest in (b).

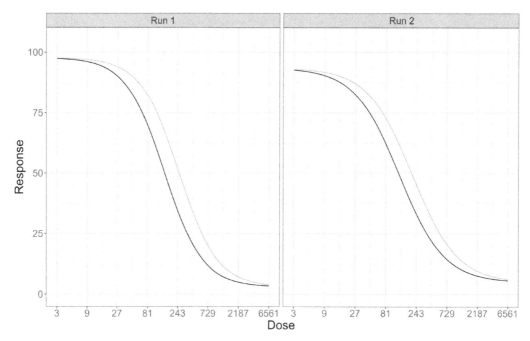

Sample — Reference standard — Test sample

FIGURE 1.5

Dose-response curves for two assay runs consisting of a reference standard and a test sample. While there are differences in the maximum and minimum responses between run 1 and run 2, the reference and test curves within the run have shifted together.

this, where it can be seen that while there are differences in the maximum and minimum responses between run 1 and run 2, the reference and test curves within the run have shifted together.

Since we can expect the factors that impact the variability of the dose-response curve to affect the reference and test samples similarly, a measurement of potency that is relative to a reference sample is likely to be less variable than one of absolute potency. Finney [19] states:

> *The ratio of two equally effective doses is an estimate of the potency of the test preparation relative to that of the standard.*

This ratio is known as relative potency (RP) and is a unitless measure of potency. The relative potency can be expressed as either a number or a percentage.

In the calculation of RP, the dose of the *test preparation*, or test sample, is conventionally in the denominator. This ensures that when a higher dose of the *standard*, or reference standard, is needed to produce the same effect, the RP for the test sample is greater than 1. For example, if two units of the reference standard and one unit of the test sample yield the same response, then the relative potency will be $2/1 = 2$ or 200%. The concentration of the test sample, in the units required for labelling, for example, can then be calculated

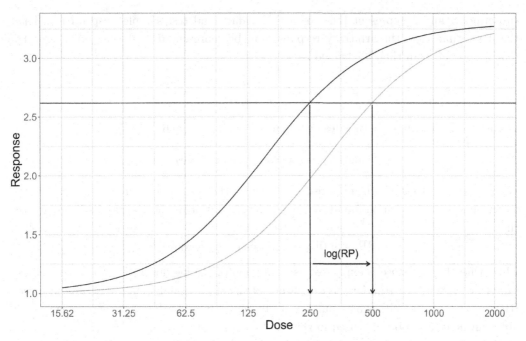

FIGURE 1.6

The relative potency is the ratio of two equally effective doses. Choosing an arbitrary response level (horizontal line) requires a dose of 250 units for the reference sample and a dose of 500 units for the test sample. Therefore, the RP = $250/500 = 0.5$ or 50%. When the dose is plotted on the log scale, the log relative potency is related to the horizontal distance between the curves.

by multiplying the estimated RP of the test sample by the (known) concentration of the reference standard.

1.3.1 Determining relative potency

The measurement of relative potency makes the implicit assumption that test samples and reference standards are identical in their biologically active component and differ only in the extent to which they are diluted by inactive materials [19]. In other words, the reference standard is biologically similar to the test sample and differs only in the concentration of the active ingredient. Therefore, the test sample behaves as a dilution of the reference standard. The relative potency then measures how much the test sample needs to be diluted (or concentrated) to behave in an identical fashion to the reference standard.

Figure 1.6 shows the dose-response relationships for a test sample and a reference standard. To determine the RP, we need to find equally effective doses for the two samples. The test sample and reference standard are *equally effective* when their responses are equal. Taking an arbitrary response (horizontal line in Figure 1.6), we can identify the doses for the two samples that result in that response, i.e., two equally effective doses. Let the subscripts S and T represent the reference standard and test sample, respectively. Then, the doses that yield the arbitrary response can be represented as dose_S and dose_T. The relative potency can then be defined as:

$$\text{RP} = \frac{\text{dose}_S}{\text{dose}_T}. \tag{1.1}$$

Taking the logarithm of both sides of Equation (1.1), the $\log(\text{RP})$ is:

$$\log(\text{RP}) = \log(\text{dose}_S) - \log(\text{dose}_T), \tag{1.2}$$

and corresponds to the horizontal distance between the two curves in Figure 1.6. Therefore, the relative potency can be found by taking the antilog of the horizontal distance between the curves:

$$\text{RP} = \text{antilog}\left[\log(\text{dose}_S) - \log(\text{dose}_T)\right]. \tag{1.3}$$

Thus, when the response is expressed as a function of the log dose, the potency of the test sample, relative to the standard, is the antilog of the horizontal distance between the curves. RP can be expressed either as a number or, by multiplying by 100, as a percentage. These are equivalent, and both are used in this book.

1.3.2   Parallelism

In the example shown in Figure 1.6, the choice of the response level was arbitrary. To ensure a uniquely defined RP, any choice of response must give the same result; i.e., the horizontal distance between the two curves is constant (see Figure 1.7). In other words, it must be possible to translate, or slide, the curve for the test sample horizontally until it sits on top of the standard curve. This property is known as parallelism. If the test sample behaves as a dilution of the standard preparation, as it should if the preparations are biologically similar, then the curves will be parallel. Note, though, that parallelism does not necessarily imply biological similarity.

If the property of parallelism does not hold, the horizontal distance between the dose-response curves will vary at different response levels (see Figure 1.8). Therefore, the RP will depend on the choice of response level and will not be unique. In this situation the RP is not well-defined and should not be reported.

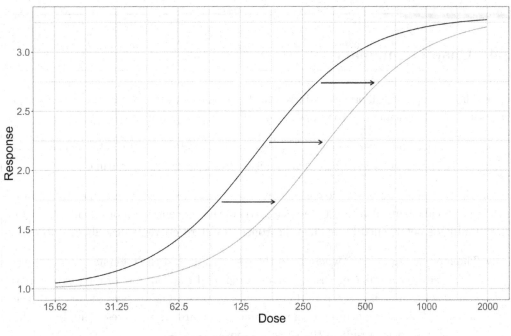

FIGURE 1.7

When the standard and test curves are parallel, the horizontal distance between the curves is constant, and the RP is independent of the response level.

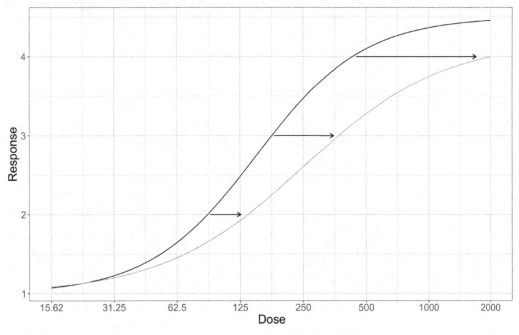

FIGURE 1.8

When the standard and test curves are non-parallel, the horizontal distance varies; the RP depends on the response level and is not uniquely defined.

1.4   Chapter summary

- The concept of a bioassay is introduced and the reasons for the need for special statistical treatment of the evaluation of a biological product are explained.

- The use a of reference standard in bioassays is introduced and the key concept of a dose-response relationship, which underlies bioassay design, is outlined.

- The mathematically simple concentration-determining assay, sometimes called bio-analytical or interpolation-type assays, where the concentration of an unknown test sample is determined by reading off from a standard curve, is described.

- The more complex case, where both the test sample and the reference standard are evaluated over a range of doses, is considered, and the concepts of relative potency and absolute potency are introduced. Relative potency is defined, in terms of the dose-response relationships of the reference standard and the test sample, as the ratio of doses that result in the same response.

- The concept of parallel line or parallel curve dose-response relationships is introduced and its fundamental importance to relative potency is explained. Effective dose, ED_{50} and effective concentration, EC_{50} are defined.

2

Bioassay performance requirements

When describing the performance characteristics of a bioassay, it is important first to define some terms. We will use the following terminology [53]:

> ***Bioassay method:*** *...operational components comprised of instrumentation, reagents, standards, sample preparations, calibrations, controls, and suitability criteria on the system.*

> ***Bioassay procedure:*** *...a use of the method to make a decision, which might detail a study design, a sampling plan, an analytical replication strategy, and calculations resulting in a reportable value.*

The bioassay method is defined as an experiment with a specified output called the measurement – a concentration or a relative potency. A single bioassay method can be associated with several procedures, typically involving different replication strategies, for example for batch release, for stability testing, or for qualification of a reference standard batch. The terms 'bioassay' and 'assay' will be used to refer to either of these, and, where necessary, we will be specific.

2.1 Purpose of a bioassay

One might state the aim of a bioassay simply as: 'to provide a measure of potency (or concentration) that is close to the true value'. But bioassays are variable and so one result for a given test sample could be very close to the true value whilst the next result might be much further away. Therefore, the aim of a bioassay might be better stated as: 'to provide a measure that has a good chance of being close to the true value'. Then, the performance of a bioassay can be captured by: (a) the average closeness of the result to the true value; and (b) the variability of the result for repeated tests of the same sample.

These characteristics of 'average closeness of the result to the true value' and 'having a good chance of being close to the true value' must be defined in terms that allow them to be objectively assessed. Then targets for these measures can be developed to determine if the new assay is fit for purpose (or not). The targets should be based on a number of factors, including clinical requirements, laboratory and throughput requirements, as well as costs.

DOI: 10.1201/9781003449195-2

Another important characteristic of a good bioassay is that it does not produce a signal for a test sample that does not contain the material of interest. This is known as specificity. Demonstrating specificity of the assay does not usually require formal statistical analysis, and we will not discuss it further.

Finally, it is important to know the set of true potency or concentration values for which the bioassay is capable of performing adequately. For relative potency assays this is called the bioassay range and is expressed in terms of relative potency values. For absolute potency or concentration, limits of quantitation are required: the lower limit of quantitation (LLOQ) and the upper limit of quantitation (ULOQ). These limits of quantitation represent the lowest and highest concentrations between which the method is suitably accurate and precise.

2.1.1   Analytical target profile

The analytical target profile (ATP) bridges the performance characteristics of the bioassay with the clinical and business requirements. The ATP for a bioassay procedure consists of a description of the intended purpose of the procedure and relevant performance character- istics with associated performance or acceptance criteria. Although the ATP applies to a bioassay procedure rather than a bioassay method, the performance criteria can be stated in terms of the bioassay method. The performance of the procedure can then be inferred from the performance of the method combined with the design of the procedure (see Section 3.4).

2.1.2   Operational considerations

Operational requirements such as high throughput and ease of use also have an impact on acceptance criteria for performance characteristics. These requirements need to be balanced with the requirements of the ATP when considering the fitness for use of the bioassay.

The remainder of this chapter considers the definition and measurement of bioassay performance characteristics and how they can be evaluated in terms of the ATP. We consider appropriate acceptance criteria, or limiting values, for the characteristics that will ensure that the assay both meets its ATP and is operationally fit for purpose. The requirements provide a basis for bioassay design: this will be discussed in Chapter 3.

2.2   Describing bioassay performance

To describe the performance of a bioassay we need to understand how the results vary from run to run. Thus, we begin by considering how to describe the distribution of results that would be seen if multiple independent runs of the bioassay method were performed for the same test sample, each resulting in a measured value. We present the concepts in terms of relative potency, RP. However, the same approach can be applied to concentrations or absolute measures of potency.

2.2.1   Distribution of repeated measurements

A distribution describes the variation in a variable quantity, such as the relative potency. The properties of the distribution can be used both to estimate the probability that the quantity will be in a given range and to perform statistical inference. We refer to Appendix A for a more detailed discussion of basic statistical methods and theory. In the statistical analysis of bioassay data it is often assumed that the values of repeated measurements, for example relative potency, follow a log normal distribution. That is, once log-transformed, the data (potency values) have a normal (bell-shaped) distribution.

When the test sample comes from a batch with a true potency or concentration that is far from zero, the distribution for the result can appear symmetric on the untransformed scale (as well as on the log-transformed scale). However, potencies and concentrations have a lower boundary of 0. Therefore, as the true potency or concentration of the test sample gets closer to this boundary, the distribution of the (raw) repeated measurements will tend to become more skewed. In this case, a log transformation of the measurements often results in normally distributed values.

We assume the log-transformed RP is normally distributed around a mean value, $\mu_{\log \mathrm{RP}}$, with variance (spread around the mean) $\sigma^2_{\log \mathrm{RP}}$. That is:

$$\log(\mathrm{RP}) \sim N\left(\mu_{\log \mathrm{RP}}, \sigma^2_{\log \mathrm{RP}}\right).$$

When the log-transformed RP measurements have mean $\mu_{\log \mathrm{RP}}$, then the geometric mean RP on the original scale will be:

$$\mathrm{Geometric\ mean(RP)} = \mathrm{antilog}(\mu_{\log \mathrm{RP}}).$$

In practice, these values will usually be unknown and will need to be estimated from data.

Figure 2.1 shows examples of the distribution of relative potencies from multiple independent runs of the assay for a given sample. The left-hand plots show the distributions of the original RP values, with geometric mean values of 0.25 and 1, respectively. On the right, we see the distributions plotted with a logged x-axis; these plots represent the probability distributions of the log(RP). When the geometric mean value is 0.25 as shown in (a), the distribution exhibits more obvious asymmetry than when the geometric mean is 1 as shown in (c). In both right-hand plots (on the log scale), the distributions are perfectly symmetric and bell-shaped.

Because relative potencies have a lower boundary of 0, observed bioassay data tend to be consistent with a log-normal distribution. An advantage of the assumption of log-normality is that it expresses the distribution of repeated values of RP in terms of only two parameters, the mean ($\mu_{\log \mathrm{RP}}$) and the variance ($\sigma^2_{\log \mathrm{RP}}$). Estimating these parameters, along with their confidence intervals, is straightforward (see Appendix A). The normal distribution also permits simple calculations of the probabilities of events, such as the probability that a bioassay result will fall in a given interval. Typically, the log transformation also results in constant values of the variance, $\sigma^2_{\log \mathrm{RP}}$, that is, the variance of logRP does not depend on the value of RP.

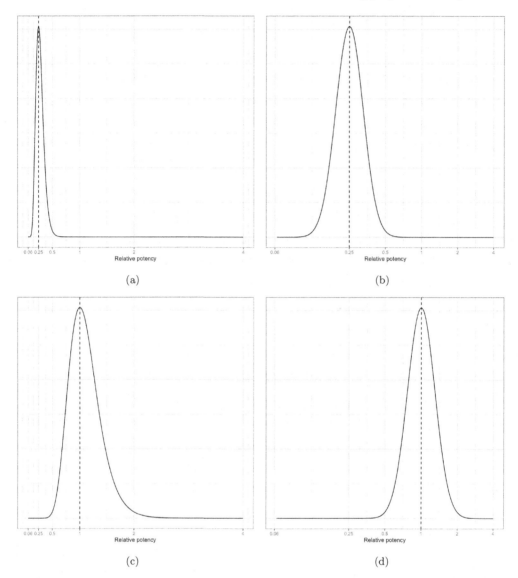

(a)   (b)

(c)   (d)

FIGURE 2.1
Frequency distributions for the RP and $\log$ RP with mean of $\mu_{\log \text{RP}} = \log(0.25)$ (geometric mean of RP = 0.25) in (a) and (b) and mean of $\mu_{\log \text{RP}} = \log(1)$ (geometric mean of RP of 1) in (c) and (d). The distribution of the raw RP values is asymmetric or skewed (plots (a) and (c)). The distribution of the log-transformed RP values is symmetric and bell-shaped (plots (b) and (d)).

If an assay is performing well, then it will have a good chance of reporting a result close to its true value. Thus, the mean of the measured results (log scale), $\mu_{\log \text{RP}}$, will be close to the true value of the RP (log scale) for the batch. In addition, the variance of the measured results (log scale), $\sigma^2_{\log \text{RP}}$, will be low. The variance will drive the design of bioassay procedures in terms of replication strategy (see discussion in Chapter 3 and Chapter 8).

In this book, we will assume that the distribution of results from multiple independent runs of a bioassay is log normal. We will use the natural logarithm, $\log_e$, or equivalently ln, throughout. All calculations will be conducted on the log scale and results back-transformed for interpretation on the original scale where appropriate.

2.3 Bioassay performance parameters

In order to assess 'the average closeness of the result to the true value' and 'having a good chance of being close to the true value' objectively, we need to define numeric measures – statistics – that can be used to quantify these characteristics. We present the concepts in terms of relative potency, RP. However, the same approach can be applied to absolute potencies or concentrations.

2.3.1 Accuracy: average closeness to true value

To assess closeness of the result to the true value, we need results for test samples with known true values. Testing the reference standard against itself is a good way to achieve this. In this case, the assay should, on average, returns a relative potency of close to 1. The closeness of the result to the true value, on average, captures systematic error in the bioassay.

Accuracy can be defined in a number of different ways. We will focus on relative accuracy (RA) and relative bias (RB). For the relative potency case, denote the true relative potency by TRP, and let RP be a measured relative potency. Then we can define:

- relative accuracy as:

$$\text{RA} = 100 \times \frac{\text{RP}}{\text{TRP}}\%; \tag{2.1}$$

- relative bias as:

$$\text{RB} = 100 \times \frac{(\text{RP} - \text{TRP})}{\text{TRP}}\%$$
$$= 100 \times \left[\frac{\text{RP}}{\text{TRP}} - 1\right]\%. \tag{2.2}$$

Note that relative accuracy is also referred to as 'recovery' and that the relative bias can also be expressed as $\text{RB} = (\text{RA} - 100)\%$. We choose RB as the preferred parameter for accuracy. In terms of the distribution of $\log(\text{RP})$, the (true) relative bias of the assay is given by:

$$\text{RB} = 100 \times \frac{\text{antilog}\left(\mu_{\log \text{RP}}\right) - \text{TRP}}{\text{TRP}}\%. \tag{2.3}$$

This can be rearranged as:

$$\frac{\text{RB}}{100} + 1 = \frac{\text{antilog}\left(\mu_{\log \text{RP}}\right)}{\text{TRP}}.$$

After taking the logarithm of both sides:

$$\log\left[\frac{\text{RB}}{100} + 1\right] = \mu_{\log \text{RP}} - \log\left(\text{TRP}\right), \tag{2.4}$$

and in the case that TRP = 1:

$$\log\left[\frac{\text{RB}}{100} + 1\right] = \mu_{\log \text{RP}}.$$

That is to say, when the true RP is 1, the mean RP on the log scale is found as the logarithm of 1 plus the RB (expressed as a proportion). The RB can be either positive or negative, with values close to zero being desirable. A positive relative bias indicates that the assay has a tendency to overestimate the TRP, while negative values indicate a tendency to underestimate the TRP. When the RB is zero, then the assay, on average, returns the TRP.

Illustration: accuracy

Consider the scenario where the reference standard is tested against itself. Then we know that the TRP is 1. However, suppose that the assay, on average, returns a result with a geometric mean RP of 0.90. Then on the log scale, $\mu_{\log \text{RP}}$, is $\log(0.9) = -0.1054$ and the distribution of the log(RP) will be centred at log(0.9). The RB will be:

$$\begin{aligned}
\text{RB} &= 100 \times \frac{\text{antilog}\left[\log\left(0.9\right)\right] - 1}{1} \\
&= 100 \times \frac{0.9 - 1}{1} \\
&= -10\%.
\end{aligned}$$

In this case, the assay tends to underestimate the true potency by 10%. Figure 2.2 shows the distribution of RP values (on the log scale) for this scenario. The bell-shaped curve shows the distribution of assay results for a test sample with true potency of 1. The centre of the curve is to the left of the true value, indicating the negative bias.

Example: estimation of accuracy

In practice, $\mu_{\log \text{RP}}$ will be unknown and will need to be estimated. Suppose a reference standard is tested (as the test sample) in a relative potency bioassay. This test sample will have the same potency as the bioassay reference standard, therefore its true relative potency TRP = 1.

A series of RP results for the reference standard as test sample provides an initial assessment of the accuracy of the bioassay, albeit limited to the case where TRP = 1. Similar experiments can be conducted for other known values of TRP. For example, the reference standard can be diluted to half strength, so that its known TRP = 0.5. It should not be assumed that the relative bias will be constant at different true potencies. In fact, it is more likely that an assay will exhibit less bias for true potencies around 100% and more bias further away. For further discussion, see Chapter 8.

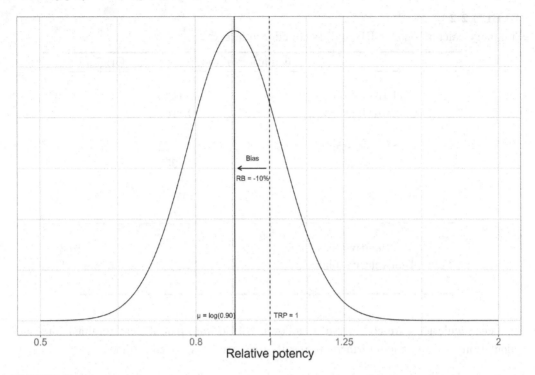

FIGURE 2.2

Distribution of RP results for a sample with $\mu_{\log RP} = \log(0.9)$. The x-axis is plotted on the log scale and the bell-shaped curve is centred at $\log(0.9)$. If the TRP = 1, the RB is -10%.

TABLE 2.1

Twelve measurements of the RP for a sample with a TRP = 1.

Observation	Result RP	Result log(RP)
1	0.95	−0.0513
2	0.84	−0.1744
3	1.06	0.0583
4	0.80	−0.2231
5	0.91	−0.0943
6	0.78	−0.2485
7	0.83	−0.1863
8	0.96	−0.0408
9	0.75	−0.2877
10	0.89	−0.1165
11	1.08	0.0770
12	0.92	−0.0834

Table 2.1 shows the results obtained from such an experiment, where 12 measurements of the RP are recorded for a sample with TRP = 1.0. Table 2.2 shows a summary of the calculations of RB as well as its two-sided 90% confidence interval (see Appendix A for formulas for mean, standard deviation and confidence intervals).

TABLE 2.2

Summary calculations for RB and its confidence interval.

	Calculations on $\log_e$ scale	Original scale
Relative potency	-0.1143	0.892
Standard deviation	0.1151	
Sample size, n	12	
$t_{0.95,11}$	1.7959	
90% CI for $\mu_{\log \mathrm{RP}}$	$(-0.1739, -0.0546)$	$(0.840, 0.947)$
Relative bias		-10.8%
90% CI for relative bias		$(-16.0\%, -5.3\%)$

The calculations are conducted as follows. All decimal places were retained for every calculation; however the intermediate results presented have been rounded.

1. The estimate of the mean RP on the (natural) log scale can be calculated as the (arithmetic) mean of the $n = 12$ observations and is $\hat{\mu}_{\log \mathrm{RP}} = -0.1143$.

2. The standard deviation of the 12 observations on the (natural) log scale is $\hat{\sigma}_{\log \mathrm{RP}} = 0.1151$.

3. The 90% confidence interval for the mean on the log scale is given by:

$$\hat{\mu}_{\log \mathrm{RP}} \pm t_{0.95, n-1} \times \frac{\hat{\sigma}_{\log \mathrm{RP}}}{\sqrt{n}}$$
$$= -0.1143 \pm 1.7959 \times \frac{0.1151}{\sqrt{12}}$$
$$= (-0.1739, -0.0546),$$

where $t_{0.95, n-1}$ is the 95th percentile of the t-distribution with $n-1$ degrees of freedom.

4. Back-transforming to the original scale by exponentiating the estimate and confidence interval for $\mu_{\log \mathrm{RP}}$, we get:

   (a) Estimate of the geometric mean RP:

$$\widehat{\mathrm{RP}} = \text{antilog}\,(-0.1143)$$
$$= 0.892;$$

   (b) 90% confidence interval for the geometric mean RP:

$$(\text{antilog}\,\{-0.1739\}, \text{antilog}\,\{-0.0546\}) = (0.840, 0.947).$$

5. Finally, applying the formula for RB (Equation (2.2)) to the estimate of the geometric mean RP and its confidence interval:

   (a) Estimate of RB:

$$\widehat{RB} = 100 \times \frac{0.892 - 1}{1}$$

$$= -10.8\%;$$

   (b) 90% confidence interval for RB:

$$\left(100 \times \frac{0.840 - 1}{1}, 100 \times \frac{0.947 - 1}{1} \right) = (-16.0\%, -5.3\%).$$

Since the confidence interval for the RB falls entirely below zero, these data are consistent with a bioassay that, on average, returns a result less than the true value. In such a case, attempts should be made to correct this via assay design features such as plate layout (see Chapter 3).

2.3.1.1 Symmetry of accuracy

Relative bias is calculated based on the ratio of the measured and true results and is symmetric on the ratio scale. Symmetry on the ratio scale implies multiplicative symmetry about a value of 1. The inaccuracy implied by a positive RB of +25% is equivalent to the inaccuracy implied by a RB of −20%. Recall:

$$RB = 100 \left(\frac{RP}{TRP} - 1 \right) \%,$$

so

$$\frac{RP}{TRP} = \frac{RB}{100} + 1.$$

Suppose that RB = 25%; then:

$$\frac{RP}{TRP} = \frac{25}{100} + 1 = 1.25.$$

The equivalent negative relative bias would be given by:

$$\frac{RP}{TRP} = \frac{1}{1.25} = 0.8,$$

so

$$RB = 100 \times (0.8 - 1) \% = -20\%.$$

A general formula for the negative relative bias (RB$^-$) equivalent of a positive relative bias RB$^+$ is:

$$RB^- = \frac{-RB^+}{\left(1 + \frac{RB^+}{100} \right)} \%. \tag{2.5}$$

TABLE 2.3

Examples of symmetric RB values on the original and log scales.

Original scale		Log scale	
Negative RB (RB$^-$)	Positive RB (RB$^+$)	$\log\left(1 + \frac{\text{RB}^-}{100}\right)$	$\log\left(1 + \frac{\text{RB}^+}{100}\right)$
-2.9%	3.0%	-0.03	0.03
-4.8%	5.0%	-0.05	0.05
-9.1%	10.0%	-0.10	0.10
-13.0%	15.0%	-0.14	0.14
-16.7%	20.0%	-0.18	0.18
-20.0%	25.0%	-0.22	0.22

This leads to additive symmetry about a value of 0 when working on the log scale. We begin by expressing RB$^-$ as a proportion by dividing both sides of Equation (2.5) by 100. We then add 1 to each side and take the logarithm to yield:

$$\log\left(\frac{\text{RB}^-}{100} + 1\right) = \log\left[\frac{-\text{RB}^+}{100\left(\frac{\text{RB}^+}{100} + 1\right)} + 1\right].$$

This can be re-expressed as:

$$\log\left(\frac{\text{RB}^-}{100} + 1\right) = \log\left[\frac{-\text{RB}^+ + 100\left(\frac{\text{RB}^+}{100} + 1\right)}{100\left(\frac{\text{RB}^+}{100} + 1\right)}\right]$$

$$= \log\left[\frac{100}{100\left(\frac{\text{RB}^+}{100} + 1\right)}\right]$$

$$= \log(1) - \log\left(\frac{\text{RB}^+}{100} + 1\right).$$

Noting that $\log(1) = 0$:

$$\log\left(\frac{\text{RB}^-}{100} + 1\right) = -\log\left(\frac{\text{RB}^+}{100} + 1\right),$$

and demonstrates symmetry about a value of 0 on the log scale.

Table 2.3 shows some examples of pairs of symmetric RB values. On the original scale, when the RB is close to zero, the equivalent positive/negative bias values are quite similar (in absolute value). However, as the RB gets further from zero, these equivalent bias values diverge (in absolute value).

2.3.2   Linearity

Linearity refers to the relationship between observed and true relative potencies. It is usually assessed on the log-log scale, comparing the measured potencies, $\log(\text{RP})$, against their corresponding true values, $\log(\text{TRP})$. If, on average, the measured potencies are equal to

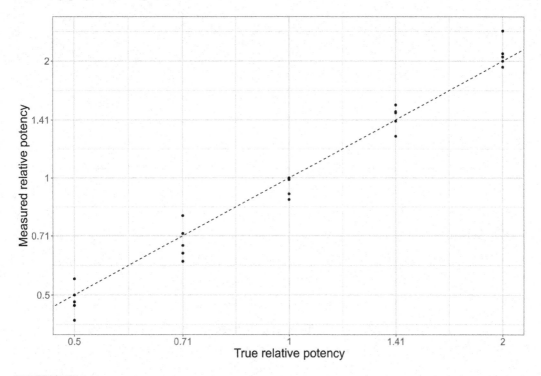

FIGURE 2.3

Plot of measured versus true potency on the log-log scale for five measurements at each of five true potencies. The line of equality is represented as a dashed line.

their true value, then, when plotted, they will be scattered about the line of equality. If this is the case, the assay is unbiased. In practice, however, the assay may exhibit some bias and an understanding of the linear relationship between measured and true potencies can help describe the nature of this bias. Figure 2.3 illustrates a plot of measured versus true potency for five measurements, at each of five true potencies, along with the line of equality.

Let $i = 1, \ldots, n$ index a set of n measured relative potencies. Then, if the relationship between an individual measured relative potency, RP_i, and its corresponding true relative potency, TRP_i, is a straight line on the log-log scale, it can be expressed as:

$$\log(\text{RP}_i) = \beta_0 + \beta_1 \times \log(\text{TRP}_i) + \epsilon_i,$$

where ϵ_i is a random error term that is normally distributed with mean 0 and variance $\sigma^2_{\log \text{RP}}$, β_0 is the y-intercept, and β_1 is the slope.

The regression line can also be stated in terms of the mean log RP for a given TRP:

$$\mu_{\log \text{RP}} = \beta_0 + \beta_1 \times \log(\text{TRP}).$$

Subtracting $\log(\text{TRP})$ from both sides:

$$\mu_{\log \text{RP}} - \log(\text{TRP}) = \beta_0 + \beta_1 \times \log(\text{TRP}) - \log(\text{TRP})$$
$$= \beta_0 + (\beta_1 - 1) \times \log(\text{TRP}).$$

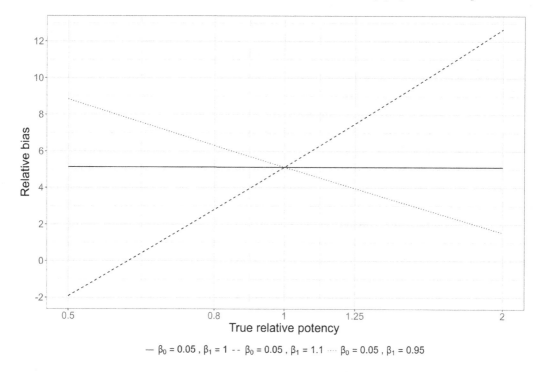

FIGURE 2.4

Plot of relative bias versus true relative potency for $\beta_0 = 0.05$ and different values of β_1. When $\beta_1 = 1$, the RB is constant at RB = 5.1%; when $\beta_1 = 1.1$, the RB increases as the TRP increases; when $\beta_1 = 0.95$, the RB decreases as the TRP increases.

Therefore, β_0 represents a measure of the bias when TRP = 1. Then from Equation (2.4):

$$\log\left(\frac{\text{RB}}{100} + 1\right) = \beta_0 + (\beta_1 - 1) \times \log\left(\text{TRP}\right).$$

By rearranging and solving for RB:

$$\text{RB} = 100 \times \{\text{antilog}\left[\beta_0 + (\beta_1 - 1) \times \log\left(\text{TRP}\right)\right] - 1\}\,\%. \tag{2.6}$$

From Equation (2.6), if, $\beta_1 = 1$ then RB is constant across all values of TRP, taking the value $100 \times [\text{antilog}(\beta_0) - 1]\,\%$. If, additionally, $\beta_0 = 0$, then RB = 0% and the assay is unbiased. If $\beta_1 \neq 1$ then RB will depend on the value of the TRP. In summary, β_0 represents the bias when TRP = 1; if $\beta_1 = 1$, then the bias is constant across values of TRP.

2.3.2.1 Proportional bias

We can express the extent of the variation in the relative bias over values of TRP by re-expressing Equation (2.6) as :

$$\text{RB} = 100 \times \left[\text{antilog}\left(\beta_0\right)\text{TRP}^{(\beta_1 - 1)} - 1\right]\,\%.$$

The factor $\text{TRP}^{(\beta_1 - 1)}$ is applied to $\text{antilog}(\beta_0)$ in the formula as shown. In this sense, the bias is proportional to a power of the TRP. Figure 2.4 demonstrates the relationship between

TABLE 2.4

RP values for 25 measurements at five true relative potency levels.

	True relative potency				
Observation	0.50	0.71	1.00	1.41	2.00
1	0.50	0.67	1.00	1.48	2.05
2	0.43	0.61	0.99	1.28	2.39
3	0.55	0.80	0.99	1.40	1.93
4	0.48	0.72	0.88	1.47	2.00
5	0.47	0.64	0.91	1.54	2.09

TABLE 2.5

Estimates for the linear regression model on the log-log scale for the measured and true relative potencies in Table 2.4.

	Estimate	Std. error	90% CI
Intercept, $\hat{\beta}_0$	−0.012	0.016	$(-0.039, 0.016)$
Slope, $\hat{\beta}_1$	1.057	0.033	$(1.001, 1.114)$

the RB and TRP for different values of β_1 at $\beta_0 = 0.05$. When $\beta_1 = 1$, the relative bias is constant across all TRP levels. When $\beta_1 < 1$, the RB decreases as the TRP increases and when $\beta_1 > 1$, the RB increases as TRP increases. For example, when $\beta_1 = 1.1$ and TRP $= 0.5$, RB $= -1.9\%$.

This approach provides a way of deciding whether it is appropriate to calculate a single combined value for the bias across potency levels. We discuss this further in Chapter 8.

Example: estimation of linearity

As for estimation of RB, which requires estimation of the parameter $\mu_{\log \text{RP}}$, the linearity parameters of β_0 and β_1 must be estimated. The estimation of these parameters requires that relative potency be measured at multiple true values. Table 2.4 contains the results of such an experiment, where five relative potency measurements were obtained at each of five true potency levels $(0.50, 0.71, 1.00, 1.41, 2.00)$ to result in a total of $n = 25$ measurements.

Figure 2.5 plots the data in Table 2.4 along with the estimated linear regression line and line of equality. The estimates of the linear regression model are found in Table 2.5. The methodology for estimating the regression line can be found in Appendix A. The aim of such a study is to demonstrate that the slope is equivalent to 1. We can see that the estimated regression line diverges from the line of equality. An equivalence interval can be set (see Section 7.1), within which the 90% confidence interval for the slope must lie. A common choice is $(0.8, 1.25)$. In this case, the 90% CI for β_1 is $(1.001, 1.114)$ which lies within the equivalence bounds and so supports the hypothesis that the slope is equivalent to 1 and implies the bias is constant across TRP.

2.3.3 Precision: variability of repeated measurements

A good assay will be one that produces results that are close to one another when testing is repeated. Variability of repeated tests captures the random component of error in the

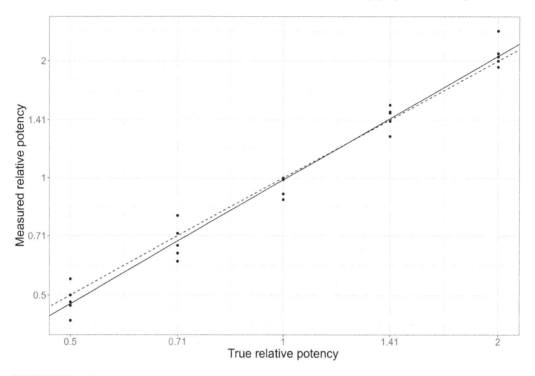

FIGURE 2.5

Plot of measured vs true potency on the log-log scale for five measurements at each of five true potencies. The estimated regression line is shown as a solid line, while the line of equality is represented as a dashed line.

bioassay. In order to assess the bioassay variability, we need results for repeated runs of the method testing the same sample (regardless of its true value).

The variability of repeated tests can be described in several ways. For example, on the observed scale we may choose to use the standard deviation, the coefficient of variation (also known as the relative standard deviation), the range or the inter-quartile range. See Appendix A for definitions of these terms. However, as the RP is log-normally distributed (Section 2.2.1), working with the data on the log scale is the obvious choice.

Recall that we assume:

$$\log\left(\text{RP}\right) \sim N\left(\mu_{\log\text{RP}}, \sigma^2_{\log\text{RP}}\right).$$

The variance, $\sigma^2_{\log\text{RP}}$, or the standard deviation, $\sqrt{\sigma^2_{\log\text{RP}}} = \sigma_{\log\text{RP}}$, can be used to express the variability. Another measure of the variability is the geometric coefficient of variation (GCV) [71], defined as:

$$\text{GCV} = 100 \times \left[\text{antilog}\left(\sqrt{\sigma^2_{\log\text{RP}}}\right) - 1\right]\%.$$

The GCV is independent of the scale of the measurement on the original scale and is also independent of the base for the logarithm. As for the variance, the smaller the value of the GCV, the less variable the observations.

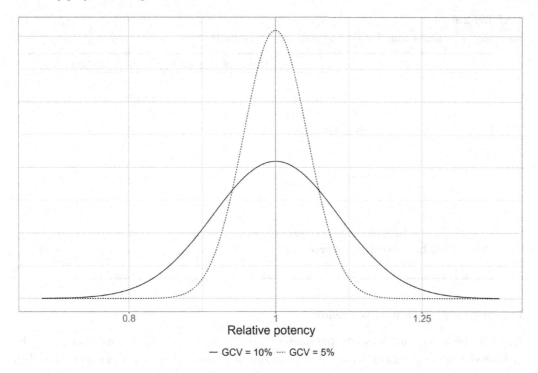

Relative potency

— GCV = 10% ···· GCV = 5%

FIGURE 2.6
Distributions for log(RP) with $\mu_{\log \mathrm{RP}} = 1$ and GCV of 5% and 10%. The smaller the GCV, the narrower the distribution and the more concentrated the values are about the mean.

Re-arranging the equation for the GCV, the antilog of the standard deviation of the log(RP) is:

$$\text{antilog}\left(\sigma_{\log \mathrm{RP}}\right) = \frac{\mathrm{GCV}}{100} + 1,$$

and, by taking the logarithm of both sides, the standard deviation of the log(RP) in terms of the GCV is:

$$\sigma_{\log \mathrm{RP}} = \log\left(\frac{\mathrm{GCV}}{100} + 1\right).$$

We will define the intermediate precision (IP) of the bioassay method as the GCV of repeated runs of the method:

$$\mathrm{IP} = 100 \times \left[\text{antilog}\left(\sqrt{\sigma_{\log \mathrm{RP}}^2}\right) - 1\right]\%.$$

Note that low values of IP are desirable.

Illustration: precision

Figure 2.6 shows the distribution of RP values (on the log scale) with GCVs of 5% and 10% for a geometric mean RP of 1. The smaller the GCV, the narrower the distribution, and the more concentrated the results about their mean. For a geometric mean RP of 1, at a GCV of 10%, approximately 95% of values will fall between 83.0% and 120.5%. However, if the GCV is 5%, then this interval will be narrower and 95% of observations will fall between 90.9% and 110.0%, instead.

TABLE 2.6

Summary of calculations for IP and its confidence limit.

	Calculations on $\log_e$ scale	Original scale
Standard deviation, $\hat{\sigma}_{\log \mathrm{RP}}$	0.1151	
Sample size, n	12	
$\chi^2_{0.05,11}$	4.5748	
95% UCL for $\sigma_{\log \mathrm{RP}}$	0.1785	
Intermediate precision		12.2%
95% UCL for intermediate precision		19.5%

Example: estimation of precision

As with the mean and linearity parameters, the variance, $\sigma^2_{\log \mathrm{RP}}$, of the assay must be estimated from experimental data. For example, suppose a batch of material is tested in many runs of the bioassay. This will provide an initial assessment of the precision of the bioassay (which may be limited to batches with similar TRP). We refer to the example data presented in Table 2.1 and consider the estimate of the variability. Table 2.6 shows a summary of the calculations for the IP, expressed as a GCV, as well as its one-sided 95% upper confidence limit. Note that for precision, interest typically lies in showing that the IP is less than some threshold and so we typically only report a one-sided upper confidence limit. Also, note that a one-sided 95% confidence limit is consistent with the two-sided 90% confidence limits calculated above for the RB.

The calculations are conducted as follows. All decimal places were retained for every calculation; however the intermediate results presented have been rounded.

1. The estimate of the standard deviation on the (natural) log scale, $\sigma_{\log \mathrm{RP}}$, is calculated as the sample standard deviation of the 12 observations, where $\hat{\sigma}_{\log \mathrm{RP}} = 0.1151$.

2. The 95% (one-sided) upper confidence limit (UCL) for the standard deviation on the log scale is [8]:

$$\mathrm{UCL}_{0.95, \sigma_{\log \mathrm{RP}}} = \hat{\sigma}_{\log \mathrm{RP}} \times \sqrt{\frac{(n-1)}{\chi^2_{0.05, n-1}}}$$

$$= 0.1151 \times \sqrt{\frac{(12-1)}{4.5748}}$$

$$= 0.1785,$$

where $\chi^2_{0.05,r}$ is the 5th percentile of the χ^2 distribution with degrees of freedom of $r = n - 1$.

3. Back-transforming to the original scale, our estimate of the IP, expressed as a GCV, is:

$$\widehat{\text{IP}} = 100 \times [\text{antilog}\,(\hat{\sigma}_{\log \text{RP}}) - 1]\,\%$$
$$= 100 \times [\text{antilog}\,(0.1151) - 1]\,\%$$
$$= 12.2\%,$$

with a upper (one-sided) 95% confidence limit for IP of:

$$\text{UCL}_{0.95,\text{IP}} = 100 \times \left[\text{antilog}\,\left(\text{UCL}_{95,\sigma_{\log \text{RP}}}\right) - 1\right]\,\%$$
$$= 100 \times [\text{antilog}\,(0.1785) - 1]\,\%$$
$$= 19.5\%.$$

If the precision needs to be improved, replication should be considered: see Chapter 3.

2.4 Combining accuracy and precision: assumed RB and IP

The accuracy of a bioassay represents how close, on average, a measurement is to the true value, while the variability of an assay represents how close the results are to one another with repeated testing. Together, accuracy and precision determine the chance that a single future measure of potency will be close to its true value, and, conversely, the chance that a single future measure of potency will fall outside the specification limits. An assay with very low RB can afford to have higher IP than one with higher RB, and vice versa. Therefore, when defining the ATP, it can be beneficial to make assumptions about the values of the RB and IP to ensure that the assay is meeting the clinical and business requirements. In this section, we explore combinations that provide an interpretation in terms of the bioassay performance.

2.4.1 Probability of an out-of-specification result

The probability that an individual measurement (from a single run of the bioassay method) will be outside a given interval, for example the interval between the specification limits, can be calculated directly from the normal distribution. Areas under the normal (bell) curve correspond to probabilities. Figure 2.7 shows the areas that correspond to the probability that the result is OOS for TRPs of 1, 0.5, and 2, in panels (a), (b), and (c), respectively, for an assay with a RB $= -10\%$.

To provide a rational basis for the ATP, it can be beneficial to consider how the combination of the RB and IP can lead to an assay that is capable of discriminating the 'good' from the 'bad' batches. That is, the probability that batches that are truly within the specification and batches that are truly outside the specification limits test correctly. For the example shown in Figure 2.7, because the RB is negative, the probability of an OOS result for TRP $= 0.5$ is larger than TRP $= 2$, despite these being equidistant from their closest specification limit.

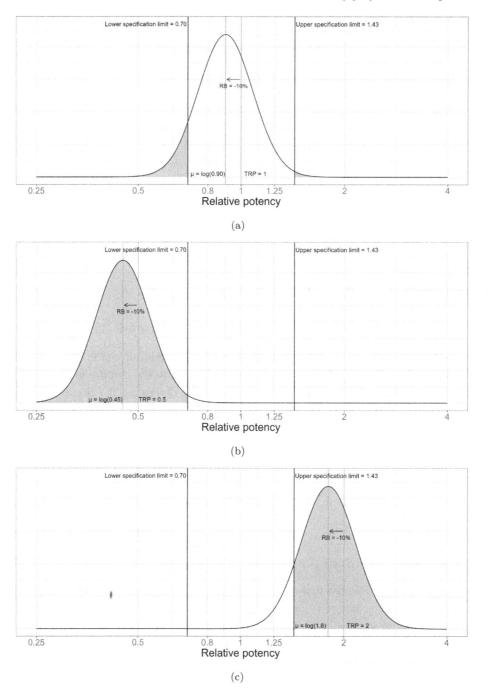

FIGURE 2.7

Distributions for the measured relative potency for samples with TRPs of 1 (inside specification limits) and 0.5 and 2 (outside specification limits), RB = −10% and IP = 20%,
in (a), (b), and (c), respectively. The specification limits are plotted as solid vertical lines.
The area corresponding to the probability of an out-of-specification RP is shaded in grey.
The relative bias of −10% results in asymmetry with respect to the symmetric specification
limits.

Suppose we have lower and upper specification limits of LSL and USL on the observed scale. Then, the probability that an individual lot will be out-of-specification (OOS) is:

$$Prob(\text{OOS}) = \left\{ \Phi \left[\frac{\log(\text{LSL}) - \mu_{\log \text{RP}}}{\sqrt{\sigma^2_{\log \text{RP}}}} \right] \right\} + \left\{ 1 - \Phi \left[\frac{\log(\text{USL}) - \mu_{\log \text{RP}}}{\sqrt{\sigma^2_{\log \text{RP}}}} \right] \right\}, \quad (2.7)$$

or equivalently, in terms of RB and IP:

$$Prob(\text{OOS}) = \left\{ \Phi \left[\frac{\log(\text{LSL}) - \log\left(\frac{\text{RB}}{100} + 1\right) - \log(\text{TRP})}{\log\left(\frac{\text{IP}}{100} + 1\right)} \right] \right\}$$
$$+ \left\{ 1 - \Phi \left[\frac{\log(\text{USL}) - \log\left(\frac{\text{RB}}{100} + 1\right) - \log(\text{TRP})}{\log\left(\frac{\text{IP}}{100} + 1\right)} \right] \right\}, \quad (2.8)$$

where $\Phi[z]$ is the cumulative distribution function of the standard normal distribution. These calculations are simply standardising the specification limits (i.e., computing a z-score) and finding the area under the standard normal curve; see Appendix A for more details.

Note that the first term in Equation 2.7 is the probability that a randomly selected result will be below the lower specification limit. The second term is the probability that a randomly selected result will be above the upper specification limit. Finally, the probability that a randomly selected result will be within the specification limits can be found by subtracting the sum of these two terms from 1.

Illustration: calculating probability of an OOS result

The effects of the RB and IP on the probability of an out-of-specification result can support the development of the ATP. They can also inform the design of procedures such as the release assay (see Chapter 3).

For specification limits of (0.70, 1.43), true potency of 1, RB = −10% and IP = 20% the probability of an out-of-specification result is:

$$Prob(\text{OOS}) = \left\{ \Phi \left[\frac{\log(0.70) - \log\left(\frac{-10}{100} + 1\right) - \log(1)}{\log\left(\frac{20}{100} + 1\right)} \right] \right\}$$
$$+ \left\{ 1 - \Phi \left[\frac{\log(1.43) - \log\left(\frac{-10}{100} + 1\right) - \log(1)}{\log\left(\frac{20}{100} + 1\right)} \right] \right\}$$
$$= \Phi \left(\frac{-0.2513}{0.1823} \right) + 1 - \Phi \left(\frac{0.4630}{0.1823} \right)$$
$$= \Phi(-1.3784) + 1 - \Phi(2.5397)$$
$$= 0.0840 + 0.0055$$
$$= 0.0896.$$

The probability of a result being below the lower specification is 0.0840, and the probability of a result being above the upper specification limit is 0.0055 to give a total probability of

TABLE 2.7

Probability of an out-of-specification result when the true potency is 1 for different combinations of RB and IP and specification limits of (0.70, 1.43).

| Relative bias | Intermediate precision | | | |
	5%	10%	15%	20%
−10%	<0.001%	0.42%	3.7%	9.0%
−5%	<0.001%	0.07%	1.6%	5.9%
0	<0.001%	0.02%	1.1%	5.0%
5%	<0.001%	0.06%	1.5%	5.8%
10%	<0.001%	0.30%	3.1%	8.2%

an out-of-specification result of 0.0896 (or 8.96%). Note that values were not rounded until the final result. This is shown in Figure 2.7(a).

For specification limits of (0.70, 1.43), Table 2.7 shows the probability of an out-of-specification result when the true potency is 1 for different combinations of RB and IP. When IP = 20%, there is a high probability that an individual batch with true potency of 1 will fall out of specification: about 6% when RB = −5%, or 8% when RB = 10%. When IP = 10%, the probability is less than 0.01, or 1%, for both RB = 5% and RB = 10%.

Figure 2.8 shows how the probability of an out-of-specification result changes with respect to the TRP of the test sample (over the range of TRP = 0.5–2.0) for various combinations of RB and IP. Figure 2.9 shows the same data for a narrower range of true potency values from 0.70 – 1.43. Reference lines at $P(\text{OOS}) = 0.01$ and $P(\text{OOS}) = 0.05$ are included.

With IP = 20% there is at least a 5% probability of an out-of-specification result, regardless of the true value for the test sample. This level of risk is unlikely to be acceptable, so the precision of the assay must be improved or the release assay procedure must include more than a single run of the method. With IP = 10%, if RB = −5% the probability of an out-of-specification result is at most 1% when the true value for the test sample is between 0.92 and 1.2; if RB = −10% the probability is at most 1% when the true value for the test sample is between between 0.80 and 1.04. Whether or not these levels of risk are acceptable will depend on the ATP.

2.4.2   Prediction interval for distance from true value

The two characteristics of RB and IP can be combined into a measure of the chance that an individual measure of potency will be within a given distance, d, of its true value [3]. The distance of the result from its true value on the log scale is:

$$d = \log(\text{RP}) - \log(\text{TRP}).$$

This corresponds to the log-transformed relative accuracy and is also a measure of bias.

For a random test of a given lot, a $100p\%$ prediction interval for the distance is an interval within which $100p\%$ of distances are expected to fall. When the mean and variance

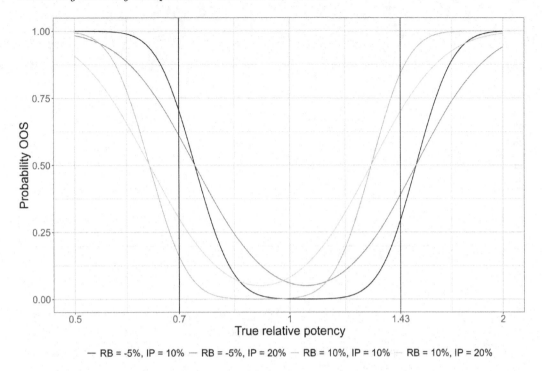

FIGURE 2.8

Probability of an out-of-specification result against true potency (range from 0.5 to 2) for different combinations of RB and IP; specification limits of (0.70, 1.43) are plotted as vertical lines. Ideally, this probability should be low for TRPs within the specification limits and high for those outside.

are assumed to be known, the $100p\%$ prediction interval for d is given by:

$$[\mu_{\log RP} - \log(\text{TRP})] \pm z_{\frac{1+p}{2}} \times \sqrt{\sigma^2_{\log RP}}, \tag{2.9}$$

or equivalently, in terms of RB and IP:

$$\log\left(\frac{\text{RB}}{100} + 1\right) \pm z_{\frac{1+p}{2}} \times \sqrt{\left[\log\left(\frac{\text{IP}}{100} + 1\right)\right]^2}, \tag{2.10}$$

where $z_{\frac{1+p}{2}}$ is the $100 \times \frac{1+p}{2}$ percentile of the standard normal distribution. If the limits are back-transformed, they become an interval for RB on a multiplicative scale. This prediction interval for the distance is centred on the bias (on the log scale) and does not provide an interval for the measured potency value itself. If an interval for the measured potency value is desired, the prediction interval can be re-centred by adding $\log(\text{TRP})$ to the left-hand side of Formula (2.10).

Illustration: prediction interval

It can be useful to explore prediction limits based on assumed values of accuracy and precision (rather than estimates from data). This approach can be valuable when exploring acceptable values for these properties, for example.

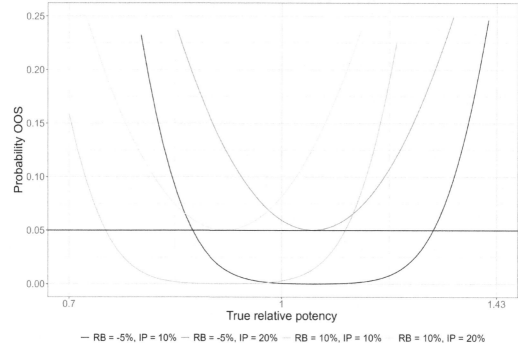

FIGURE 2.9

Probability of an out-of-specification result against true potency (range from 0.7 to 1.43) for different combinations of RB and IP and specification limits of (0.70, 1.43).

The 99% prediction interval for values of the distance of the result from the true value when the accuracy and precision are both known at RB $= -5\%$ and IP $= 10\%$ can be found as:

$$\log\left(\frac{-5}{100}+1\right) \pm z_{\frac{1+p}{2}} \times \sqrt{\left[\log\left(\frac{10}{100}+1\right)\right]^2}$$

$$= \log\left(0.95\right) \pm 2.5758 \times \log\left(1.1\right)$$

$$= -0.0513 \pm 0.2455$$

$$= \left(-0.2968, 0.1942\right),$$

and can be back-transformed to the original scale as:

$$\left(\text{antilog}\left\{-0.2968\right\}, \text{antilog}\left\{0.1942\right\}\right) = \left(0.743, 1.214\right).$$

Table 2.8 summarises this calculation. An interpretation of this interval is that 99% of log(RP) results are expected to fall within -0.2968 and $+0.1942$ of the true value, log(TRP). Back-transforming the interval represents multiples of the true value, TRP, on the original scale. Thus, 99% of RP results are expected to fall within 74.3% and 121.4% of their true potency.

Assuming the true relative potency is 1, the prediction limits for the distance are equivalent to the prediction limits for the measured RP. Figure 2.10 superimposes the prediction

TABLE 2.8

Summary of calculations for the 99% prediction interval for distance from the true value when the RB = −5% and IP = 10%.

	Calculations on $\log_e$ scale	Original scale
Relative bias		−5.0%
Intermediate precision		10.0%
Distance from true value	−0.0513	0.950
Standard deviation	0.0953	
$z_{0.995}$	2.5758	
99% Prediction interval	$(-0.2968, 0.1942)$	$(0.743, 1.214)$

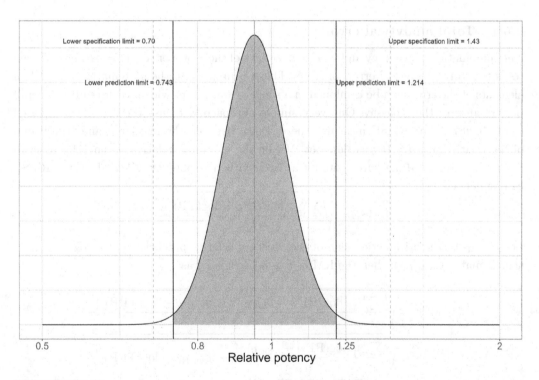

FIGURE 2.10

99% prediction limits for the distance from the true value when the RB = −5% and the IP = 10%. If the TRP = 1, then the distance is the same as the measured relative potency (shown here).

limits on the distribution for the measured RP assuming a TRP of 1. In this case, the 99% prediction interval falls within the specification limits, so we can say that there is at least a 99% chance that a single new value will fall within specification.

2.5   Combining accuracy and precision: RB and IP estimated from data

We have discussed how it can be beneficial to explore the probability of an out-of-specification result and prediction intervals based on assumed values for the accuracy and precision. However, we often wish to understand the capability of an assay and so will need to evaluate this probability based on experimental data. In this section, we assume estimates are based on a set of n measurements at a given true potency. Estimation of the accuracy and precision introduces uncertainty, and so the estimates for the prediction interval and the probability of an out-of-specification result need to take this into account. Confidence intervals quantify the uncertainty of the estimates.

2.5.1   Total analytical error

The combination of accuracy and precision (or lack of these properties) is sometimes referred to as the 'total analytical error' or TAE. The two properties, the systematic and random elements of the error, can be combined in a range of ways to provide measures of the overall performance of the bioassay. One very simple approach is to use the root-mean-squared error (RMSE). The RMSE measures the distance between the measured and true values of the relative potency (log scale). Let RP_i be the measured relative potency for a set of $i = 1, \ldots, n$ repeated measurements for a batch with true potency TRP. Then, the RMSE can be defined as:

$$\mathrm{RMSE} = \sqrt{\frac{\sum_{i=1}^{n} \left[\log(\mathrm{RP}_i) - \log(\mathrm{TRP})\right]^2}{n}}.$$

Let $\hat{\mu}_{\log \mathrm{RP}}$ be the arithmetic mean of the observed relative potencies on the log scale (i.e., the estimate of $\mu_{\log \mathrm{RP}}$); then the RMSE can be expressed as:

$$\mathrm{RMSE} = \sqrt{\frac{\sum_{i=1}^{n} \left\{[\log(\mathrm{RP}_i) - \hat{\mu}_{\log \mathrm{RP}}] + [\hat{\mu}_{\log \mathrm{RP}} - \log(\mathrm{TRP})]\right\}^2}{n}}$$

$$= \sqrt{\frac{\sum_{i=1}^{n} \left[\log(\mathrm{RP}_i) - \hat{\mu}_{\log \mathrm{RP}}\right]^2}{n} + [\hat{\mu}_{\log \mathrm{RP}} - \log(\mathrm{TRP})]^2}$$

$$\approx \sqrt{\hat{\sigma}_{\log \mathrm{RP}}^2 + \left[\log\left(\frac{\widehat{\mathrm{RA}}}{100}\right)\right]^2}.$$

Note: this is a large-sample approximation because $\hat{\sigma}_{\log\mathrm{RP}}^2 = \frac{\sum_{i=1}^n [\log(\mathrm{RP}_i) - \hat{\mu}_{\log\mathrm{RP}}]^2}{n-1}$. The RMSE can also be written in terms of RB and IP:

$$\mathrm{RMSE} \approx \sqrt{\left[\log\left(\frac{\widehat{\mathrm{IP}}}{100} + 1\right)\right]^2 + \left[\log\left(\frac{\widehat{\mathrm{RB}}}{100} + 1\right)\right]^2}.$$

Although this measure of TAE captures both elements of performance, it is not, on its own, easily interpreted. Rather, we recommend evaluating the probability of an out-of-specification result instead.

2.5.2   Probability of an out-of-specification result

The probability that an individual result will be outside (or inside) a given interval can be calculated directly when RB and IP are estimated from data.

Suppose we have lower and upper specification limits of LSL and USL (original scale). The probability that an individual lot will be out of specification is estimated by [37]:

$$\widehat{Prob}(\mathrm{OOS}) = \Phi\left[\frac{\log(\mathrm{LSL}) - \hat{\mu}_{\log\mathrm{RP}}}{\sqrt{\hat{\sigma}_{\log\mathrm{RP}}^2 \times \frac{n-1}{n}}}\right] + 1 - \Phi\left[\frac{\log(\mathrm{USL}) - \hat{\mu}_{\log\mathrm{RP}}}{\sqrt{\hat{\sigma}_{\log\mathrm{RP}}^2 \times \frac{n-1}{n}}}\right], \qquad (2.11)$$

or equivalently, in terms of RB and IP:

$$\widehat{Prob}(\mathrm{OOS}) = \Phi\left[\frac{\log(\mathrm{LSL}) - \log\left(\frac{\widehat{\mathrm{RB}}}{100} + 1\right) - \log(\mathrm{TRP})}{\log\left(\frac{\widehat{\mathrm{IP}}}{100} + 1\right) \times \sqrt{\frac{n-1}{n}}}\right]$$

$$+ 1 - \Phi\left[\frac{\log(\mathrm{USL}) - \log\left(\frac{\widehat{\mathrm{RB}}}{100} + 1\right) - \log(\mathrm{TRP})}{\log\left(\frac{\widehat{\mathrm{IP}}}{100} + 1\right) \times \sqrt{\frac{n-1}{n}}}\right], \qquad (2.12)$$

where $\Phi[z]$ is the cumulative distribution function of the standard normal distribution. Note Equation (2.11) is similar to Equation (2.7). It can be obtained by replacing $\mu_{\log\mathrm{RP}}$ and $\sigma_{\log\mathrm{RP}}^2$ by their estimates and introducing a factor of $\sqrt{\frac{n-1}{n}}$ in the denominator.

A one-sided $(1-\alpha) \times 100\%$ upper confidence limit for the probability estimate is given by the following series of calculations [37]:

1. Calculate:

$$K_L = \frac{[\hat{\mu}_{\log\mathrm{RP}} - \log(\mathrm{LSL})]}{\hat{\sigma}_{\log\mathrm{RP}}}, \qquad K_U = \frac{[\log(\mathrm{USL}) - \hat{\mu}_{\log\mathrm{RP}}]}{\hat{\sigma}_{\log\mathrm{RP}}}$$

   – If $\max(K_L, K_U) > \frac{(n-1)}{\sqrt{n}}$, set $K^* = \min(K_L, K_U)$; go to Step 5.

   – If $\max(K_L, K_U) \leq \frac{(n-1)}{\sqrt{n}}$, go to Step 2.

2. Calculate $\hat{\pi} = \widehat{Prob}(\mathrm{OOS})$ as above.

3. Determine b, the $100\widehat{\pi}$ percentile of the beta distribution with shape parameters $\frac{n-2}{2}, \frac{n-2}{2}$.

4. Calculate $K^* = \frac{(1-2b)(n-1)}{\sqrt{n}}$.

5. Solve for λ: $Prob\left(t_{\lambda;n-1} < \sqrt{n} \times K^*\right) = 1 - \alpha$, where $t_{\lambda;n-1}$ is a random variable with a non-central t-distribution with non-centrality parameter λ and $n-1$ degrees of freedom.

6. Finally, the upper confidence limit for $Prob$(OOS) is given by:

$$U = 1 - \Phi\left(\frac{\lambda}{\sqrt{n}}\right).$$

Example: probability of an out-of-specification result

When $\widehat{RB} = -10.8\%$ and $\widehat{IP} = 12.2\%$ (as estimated from the data presented in Table 2.1 with 12 observations, and rounded to 1 decimal place), for specification limits of $(0.70, 1.43)$ and TRP = 1, the estimated probability that a new single RP result will be OOS is 1.39%. This can be calculated as follows:

$$\widehat{Prob(\text{OOS})} = \Phi\left[\frac{\log(0.70) - \log\left(\frac{-10.8}{100} + 1\right) - \log(1)}{\log\left(\frac{12.2}{100} + 1\right) \times \sqrt{\frac{12-1}{12}}}\right]$$

$$+ 1 - \Phi\left[\frac{\log(1.43) - \log\left(\frac{-10.8}{100} + 1\right) - \log(1)}{\log\left(\frac{12.2}{100} + 1\right) \times \sqrt{\frac{12-1}{12}}}\right],$$

$$= \Phi\left(\frac{-0.2424}{0.1102}\right) + 1 - \Phi\left(\frac{0.4720}{0.1102}\right)$$

$$= \Phi\left(-2.1993\right) + 1 - \Phi\left(4.2823\right)$$

$$= 0.0139 + 0.0000$$

$$= 0.0139.$$

In this case, all of the estimated probability that a result is OOS is associated with being below the lower specification limit.

For the upper 95% (one-sided) confidence limit:

1. Calculate:

$$K_L = \frac{[\hat{\mu}_{\log \text{RP}} - \log(\text{LSL})]}{\hat{\sigma}_{\log \text{RP}}}$$

$$= \frac{[\log(0.892) - \log(0.70)]}{\log(1.122)}$$

$$= 2.106$$

$$K_U = \frac{[\log(\text{USL}) - \hat{\mu}_{\log \text{RP}}]}{\hat{\sigma}_{\log \text{RP}}}$$

$$= \frac{[\log(1.43) - \log(0.892)]}{\log(1.122)}$$

$$= 4.100$$

Therefore, $\max(K_L, K_U) = 4.100 > \frac{(n-1)}{\sqrt{n}} = 3.175$.

Set: $K^* = \min(K_L, K_U) = 2.106$.

5. Find:

$$\lambda : Prob\left(t_{\lambda;11} < \sqrt{12} \times 2.106\right) = 0.95$$

$$\lambda = 4.186.$$

6. Finally:

$$U = 1 - \Phi\left(\frac{4.186}{\sqrt{12}}\right)$$

$$= 11.35\%.$$

Note that in Step 1, $\max(K_L, K_U) > \frac{(n-1)}{\sqrt{n}}$ and so we skipped Steps 2–4. The upper (one-sided) 95% confidence limit for the probability that a new single RP result will be OOS is 11.35%. This limit is a lot higher than the estimate of 1.39%. This is because the estimate is based on relatively few ($n = 12$) observations.

2.5.3 Process capability index

Let $i = 1, \ldots, n$ index a set of replicate measurements of potency. A measure that is sometimes used to express the performance of a bioassay is the process capability index [10], C_{pm}, defined as:

$$C_{pm} = \frac{\log(\text{USL}) - \log(\text{LSL})}{6 \times \sqrt{\frac{\sum_{i=1}^{n}[\log(\text{RP}_i) - \log(\text{TRP}_i)]^2}{n}}}.$$

The C_{pm} is a function of the RMSE defined in Section 2.5.1 and can be written (for large n) as

$$C_{pm} = \frac{\log(\text{USL}) - \log(\text{LSL})}{6 \times \sqrt{\left[\log\left(\frac{\text{IP}}{100} + 1\right)\right]^2 + \left[\log\left(\frac{\text{RB}}{100} + 1\right)\right]^2}}.$$

When RB = 0 and the specification limits are symmetrical around 1 (multiplicatively), i.e., $\log(\text{LSL}) = -\log(\text{USL})$, the probability of an out-of-specification result is given by:

$$Prob(\text{OOS}) = 2 \times \Phi(-3 \times C_{pm}). \tag{2.13}$$

As the value of the C_{pm} increases, the probability of an out-of-specification result decreases. Consequently, large values of the C_{pm} are desirable.

TABLE 2.9

C_{pm} for specification limits of (0.70, 1.43) at various combinations of RB and IP.

Relative bias	Intermediate precision			
	5%	10%	15%	20%
−10%	1.025	0.838	0.680	0.565
−5%	1.682	1.100	0.800	0.629
0	2.440	1.249	0.852	0.653
5%	1.725	1.112	0.804	0.631
10%	1.112	0.883	0.704	0.579

TABLE 2.10

Approximate probability of an out-of-specification result based on the C_{pm} for specification limits of (0.70, 1.43) at various combinations of RB and IP.

Relative bias	Intermediate precision			
	5%	10%	15%	20%
−10%	0.21%	1.19%	4.13%	8.99%
−5%	0.00%	0.10%	1.64%	5.93%
0	0.00%	0.02%	1.06%	5.01%
5%	0.00%	0.09%	1.58%	5.84%
10%	0.09%	0.81%	3.47%	8.25%

When RB $\neq 0$, or the specification limits are not symmetric around 1, Equation (2.13) for *Prob*(OOS) does not hold and any calculated probabilities will only be approximate. Therefore, direct calculation of *Prob*(OOS), as shown in Section 2.4.1, is recommended.

Table 2.9 records the C_{pm} for specification limits of (0.70, 1.43) for various combinations of RB and IP; Table 2.10 records the approximate probability of an out-of-specification result for the same specification limits and range of RB and IP values. It can be seen that as the RB gets closer to 0, the value of the C_{pm} increases, and the approximate probability of an out-of-specification result decreases. As the IP increases, the value of the C_{pm} decreases and is associated with higher probability of an out-of-specification result.

2.5.4   Prediction interval

The prediction interval for the distance from the true value can be estimated from a set of n data values as:

$$[\widehat{\mu}_{\log \mathrm{RP}} - \log(\mathrm{TRP})] \pm t_{\frac{(1+p)}{2},(n-1)} \times \sqrt{\widehat{\sigma}^2_{\log \mathrm{RP}} \times \left(1 + \frac{1}{n}\right)}$$

where $\widehat{\mu}_{\log \mathrm{RP}}$ and $\widehat{\sigma}^2_{\log \mathrm{RP}}$ are the estimates of the mean and variance, and $t_{\frac{(1+p)}{2};(n-1)}$ is the $100 \times \frac{(1+p)}{2}$ percentile of the t distribution with $n-1$ degrees of freedom. When compared to Equations (2.9) and (2.10), the additional term under the square root of $\left(1 + \frac{1}{n}\right)$ accounts for the uncertainty in the value of $\mu_{\log \mathrm{RP}}$. Replacing the standard normal quantile z by a quantile from the t distribution accounts for the uncertainty in estimating $\sigma^2_{\log \mathrm{RP}}$.

In terms of estimates of RB and IP, denoted by $\widehat{\text{RB}}$ and $\widehat{\text{IP}}$, the prediction interval is given by:

$$\log\left(\frac{\widehat{\text{RB}}}{100}+1\right) \pm t_{\frac{(1+p)}{2},(n-1)} \times \sqrt{\left[\log\left(\frac{\widehat{\text{IP}}}{100}+1\right)\right]^2 \times \left(1+\frac{1}{n}\right)}$$

Note that these prediction intervals are for log-transformed results. Back-transforming the limits provides a prediction interval on the original scale, expressed in terms of percentages of the true value. For confidence limits on the prediction interval we refer to the tolerance interval (see Section 2.5.5).

Example: prediction interval for distance from true value

For the data from Table 2.1, the rounded estimates of accuracy and precision are $\widehat{\text{RB}} = -10.8\%$, $\widehat{\text{IP}} = 12.2\%$. The 99% prediction interval for the distance of a test result for the given lot from its true value (log scale) is as follows:

$$\log\left(\frac{-10.8}{100}+1\right) \pm t_{\frac{(1+0.99)}{2},11} \times \sqrt{\left[\log\left(\frac{12.2}{100}+1\right)\right]^2 \times \left(1+\frac{1}{12}\right)}$$

$$= -0.1143 \pm 3.1058 \times \sqrt{[\log(1.122)]^2 \times \frac{13}{12}}$$

$$= -0.1143 \pm 3.1058 \times 0.1198$$

$$= -0.1143 \pm 0.3721$$

$$= (-0.4864, 0.2578).$$

Table 2.11 summarises the calculations for the 99% prediction interval for the distance from true value based. Based on these, 99% of values of the test results for the given lot are predicted to fall within the following distances of the true value, on the log scale: $(-0.4864, 0.2578)$. Back-transforming, 99% of all test results for the given lot are predicted to fall between 61.5% and 129.4% of the true value, on the original scale. Conversely, if the TRP is 1, because the interval does not fall within $(0.70, 1.43)$, the probability that a new result will fall inside a specification of $(0.70, 1.43)$ is less than 99%.

2.5.5   Tolerance interval

The estimated $100p\%$ prediction interval described above was for the distance from the true value and may or may not actually include $100p\%$ of results. An interval based on a sample can be constructed to include at least a proportion p of the sampled population with confidence $(1-\alpha)$. Such an interval is usually referred to as a p-content, $(1-\alpha)$ coverage tolerance interval (TI) [30].

A tolerance interval for the bioassay result is given by:

$$\widehat{\mu}_{\log\text{RP}} \pm z_{\frac{(1+p)}{2}} \times \sqrt{\frac{\widehat{\sigma}^2_{\log\text{RP}} \times \left(1+\frac{1}{n}\right) \times (n-1)}{\chi^2_{\alpha,n-1}}}.$$

TABLE 2.11

Calculations for the 99% prediction interval for the distance from true value based on data.

	Calculations on $\log_e$ scale	Original scale
Relative bias		-10.8%
Intermediate precision		12.2%
Distance from true value	-0.1143	0.892
Standard deviation	0.1151	
Sample size, n	12	
$t_{0.995}$	3.1058	
99% Prediction interval	$(-0.4864, 0.2578)$	$(0.615, 1.294)$

In terms of estimates of RB and IP, denoted by $\widehat{\text{RB}}$ and $\widehat{\text{IP}}$, the tolerance interval is given by:

$$\log(\text{TRP}) + \log\left(\frac{\widehat{\text{RB}}}{100} + 1\right) \pm z_{\frac{(1+p)}{2}} \times \sqrt{\frac{\left[\log\left(\frac{\widehat{\text{IP}}}{100} + 1\right)\right]^2 \times \left(1 + \frac{1}{n}\right) \times (n-1)}{\chi^2_{\alpha, n-1}}}.$$

In essence, the TI is similar to the prediction interval (Section 2.5.4) but replaces the estimate of the variance with its upper $(1 - \alpha)$ confidence limit and is instead centred on the mean measured potency. The tolerance interval provides an interval for the measured potency values. Consequently, a comparison with the specification limits can be made more readily for the tolerance interval than for the prediction interval.

Basing decisions on (i) whether the $(p, 1 - \alpha)$ tolerance interval falls within the specification or (ii) whether the upper one-sided $(1 - \alpha)$ confidence limit for the probability of an OOS result is at most $(1 - p)$ have been shown to be approximately equivalent [7].

Example: tolerance interval for a bioassay result

For the data presented in Table 2.1, with a TRP of 1, with rounded estimates of $\widehat{\text{RB}} = -10.8\%$ and $\widehat{\text{IP}} = 12.2\%$, the 99%/95% tolerance interval can be calculated as:

$$\log(1) + \log\left(\frac{-10.8}{100} + 1\right) \pm z_{\frac{(1+0.99)}{2}} \times \sqrt{\frac{\left[\log\left(\frac{12.2}{100} + 1\right)\right]^2 \times \left(1 + \frac{1}{12}\right) \times (12 - 1)}{\chi^2_{0.05, 12-1}}}$$

$$= -0.1143 \pm 2.5758 \times \sqrt{\frac{0.1579}{4.5748}}$$

$$= -0.1143 \pm 0.4786$$

$$= (-0.5928, 0.3643).$$

TABLE 2.12

Calculations for the 99%/95% tolerance interval for the reportable result.

	Calculations on $\log_e$ scale	Original scale
Relative bias		−10.8%
Intermediate precision		12.2%
Relative potency	−0.1143	0.892
Standard deviation	0.1151	
Sample size, n	12	
$z_{0.995}$	2.5758	
$\chi^2_{0.95,11}$	4.5748	
99%/95% tolerance interval	$(-0.5928, 0.3643)$	$(0.553, 1.439)$

Back-transforming to the original scale:

$$(\text{antilog}\,\{-0.5928\}, \ \text{antilog}\,\{0.3643\}) = (0.553, 1.439).$$

Table 2.12 summarises the calculations for the 99%/95% tolerance interval, and Figure 2.11 superimposes the tolerance interval limits on the distribution for the relative potency. Both tolerance limits fall outside the specification limits which implies that more than 1% of the reported results are expected to fall OOS, with 95% confidence.

2.6 Incorporating the manufacturing distribution

So far, we have discussed assay performance in terms of repeated tests for a given lot with a given true potency value, that is, conditional on the true value for the lot. We now consider single test results for a series of lots from a manufacturing distribution, each of which will have a different true value.

The distribution of such measured values involves both the distribution of true values from the manufacturing process and the distribution of bioassay-measured values for a given true value. Combining these can be thought of as an exercise in Bayesian statistics, in the sense that we have a conditional distribution of values, combined with the distribution of the variable that was conditioned upon.

Suppose the manufacturing process produces lots whose TRP values follow a log normal distribution; that is, $\log(\text{TRP})$ is normally distributed with mean and variance, μ_{Product} and $\sigma^2_{\text{Product}}$. Assuming RB and IP are constant across values of TRP, the distribution of

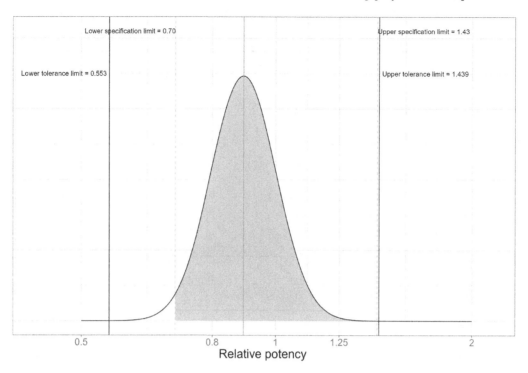

FIGURE 2.11

99%/95% tolerance limits for the RP for estimated values of RB $= -10.8\%$ and IP $= 12.2\%$. The shaded area represents the proportion of results that fall within specifciation. Since the tolerance limits fall outside the specification limits, the chance of a new value falling within specification is estimated to be less than 99%, with 95% confidence.

measured RP values for lots from the manufacturing distribution is then log normal:

$$\log(\text{RP})_{\text{Manuf}} \sim N\left(\mu_{\text{Manuf}}, \sigma^2_{\text{Manuf}}\right),$$

where:

$$\mu_{\text{Manuf}} = \mu_{\text{Product}} + \log\left(\frac{\text{RB}}{100} + 1\right), \tag{2.14}$$

and

$$\sigma^2_{\text{Manuf}} = \sigma^2_{\text{Product}} + \left[\log\left(\frac{\text{IP}}{100} + 1\right)\right]^2. \tag{2.15}$$

Figure 2.12 illustrates the manufacturing distribution of TRP values, which we cannot observe, and the distribution of bioassay-measured RP values for the manufacturing distribution, which we observe. For an unbiased assay and a manufacturing process centred at RP $= 1$, these distributions are centred at 1. The distribution for the bioassay measurements is wider as it also reflects the variability due to measurement error.

2.6.1 Proportion of out-of-specification results for manufactured lots

For the manufacturer, the proportion of manufactured lots for which the bioassay-measured result falls out-of-specification is key. This drives the ATP. For the consumer/

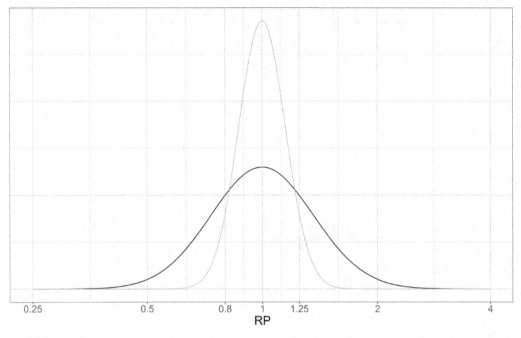

FIGURE 2.12

Distributions for the true potency of manufactured lots and the distribution for the bioassay-measured potencies. For an unbiased assay and a manufacturing process centred at RP = 1, these distributions are centred at 1. The distribution for the bioassay measurements is wider as it also reflects the variability due to measurement error.

patient/regulator, the important consideration is that lots that are in fact OOS are not released. The assessment of this aspect of assay performance can be evaluated using the methods in Section 2.5.2 for TRP values outside the specification limits.

Figure 2.13 illustrates the proportion of lots that fall out of specification via the shaded areas. The grey bell curve represents the true RP values of manufactured lots. These values are unknown but are measured using the bioassay. The black curve represents the distribution of the bioassay meaurements for the manufactured lots. The light shaded areas represent out-of-specification bioassay results.

The proportion of OOS results can be calculated using the properties of the normal distribution, see Equation (2.7), by replacing the mean and variance with the properties of the manufacturing distribution (Equations (2.14) and (2.15)) as:

$$Prob(\text{OOS}) = \Phi\left[\frac{\log(\text{LSL}) - \mu_{\text{Manuf}}}{\sqrt{\sigma^2_{\text{Manuf}}}}\right] + 1 - \Phi\left[\frac{\log(\text{USL}) - \mu_{\text{Manuf}}}{\sqrt{\sigma^2_{\text{Manuf}}}}\right], \qquad (2.16)$$

where $\Phi[z]$ is the cumulative distribution function of the standard normal distribution.

In general, μ_{Product} and $\sigma^2_{\text{Product}}$ will be unknown, at least early in the development of the bioassay. Therefore, assumptions must be made in order to calculate the proportion of

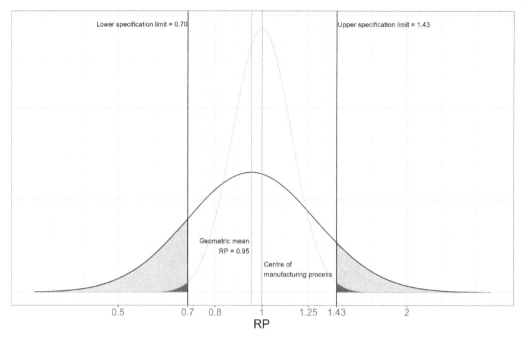

FIGURE 2.13

Out-of-specification proportions for manufucatured lots. The grey bell curve represents the distribution of the true RP values of manufactured lots. The black bell curve represents the distribution of the bioassay meaurements for the manufactured lots. The light grey shaded area represents the proportion of results that test OOS. The dark shaded area represents the proportion of truly OOS lots.

out-of-specification results and make use of it in evaluating the bioassay performance. A common assumption is that $\mu_{\text{Product}} = 0$, that is, the process is centred at $\log(\text{RP}) = 0$, or $\text{RP} = 1$. Therefore:

$$\mu_{\text{Manuf}} = \log\left[\frac{\text{RB}}{100} + 1\right].$$

A second assumption is that $\sigma^2_{\text{Product}}$ makes up a certain proportion, P, of σ^2_{Manuf}, typically about 20% [7]. The variability of the bioassay contributes the remaining proportion, $(1-P)$. Therefore:

$$\sigma^2_{\text{Product}} = P \times \sigma^2_{\text{Manuf}}.$$

Substituting this into Equation 2.15:

$$\sigma^2_{\text{Manuf}} = P \times \sigma^2_{\text{Manuf}} + \left[\log\left(\frac{\text{IP}}{100} + 1\right)\right]^2$$

$$= \frac{\left[\log\left(\frac{\text{IP}}{100} + 1\right)\right]^2}{(1-P)}.$$

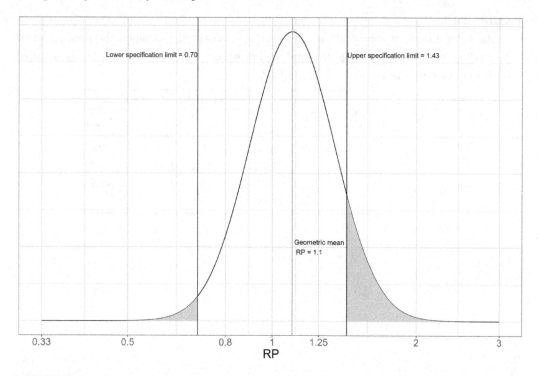

FIGURE 2.14

Proportion of out-of-specification results for manufactured lots with a process centre of 1, product variance $= 0.2$ times total variance, relative bias of 10%, intermediate precision of 20% and specification limits of (0.70, 1.43). The shaded area of 11.23% represents the proportion of out of specification results. Since there is a positive bias, the area above the upper specification limit is greater than the area below the lower limit.

Therefore:

$$Prob(\text{OOS}) = \Phi \left[\frac{\log(\text{LSL}) - \log\left(\frac{\text{RB}}{100} + 1\right)}{\frac{\log\left(\frac{\text{IP}}{100} + 1\right)}{\sqrt{1 - P}}} \right]$$

$$+ 1 - \Phi \left[\frac{\log(\text{USL}) - \log\left(\frac{\text{RB}}{100} + 1\right)}{\frac{\log\left(\frac{\text{IP}}{100} + 1\right)}{\sqrt{1 - P}}} \right].$$

Illustration: proportion of out-of-specification results for manufactured lots

Table 2.13 shows the proportion of an out-of-specification results, for various assumed values of RB and IP, for manufactured lots when the product mean is 1, product variance is 20% of the total variance and with specification limits of (0.70, 1.43). When IP $= 20\%$, there is a good chance (about 9% when RB $= -5\%$, or 11% when RB $= 10\%$) that an individual lot with true potency of 1 will fall out of specification. When IP $= 10\%$, the probability is less than 0.01, or 1%, in both cases. Figure 2.14 shows the distribution for the case when IP $= 20\%$ and RB $= 10\%$. Since there is a positive bias, the area above the upper specification limit is greater than the area below the lower limit.

TABLE 2.13

Probability of an out-of-specification result for manufactured lots when the product mean is 1, product variance is 20% of the total variance and with specification limits of (0.70, 1.43) for various assumed values of RB and IP.

Relative bias	Intermediate precision			
	5%	10%	15%	20%
−10%	<0.001%	0.92%	5.54%	12.04%
−5%	<0.001%	0.21%	2.98%	8.95%
0	<0.001%	0.08%	2.23%	7.97%
5%	<0.001%	0.19%	2.88%	8.82%
10%	<0.001%	0.69%	4.85%	11.23%

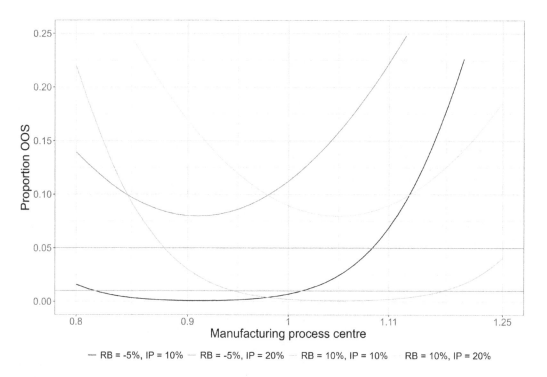

— RB = -5%, IP = 10%   — RB = -5%, IP = 20%   — RB = 10%, IP = 10%   RB = 10%, IP = 20%

FIGURE 2.15

Probability of an out-of-specification result for a range of manufacturing process centres, for RB=-5% or RB=10% and IP=10% or IP=20% for symmetric specification limits of (0.7, 1.43). Horizontal lines at 0.01 and 0.05 are shown to aid interpretation. For large values of IP (20%), the proportion OOS never falls below 5%.

Figure 2.15 shows the probability of an out-of-specification result for a test sample with a given true potency (x axis) for a range of manufacturing process centres, for combinations of RB = −5% and 10% at IP = 10% and 20%. With IP = 20% the proportion of out-of-specification results is at least 8% regardless of the centre of the manufacturing process. This level of risk is unlikely to be acceptable, so the precision of the assay must be improved or the release assay procedure must include more than a single run of the method. With IP = 10%, if RB = −5% the proportion of out-of-specification results is at most 1% so long as the manufacturing process mean is between 0.95 and 1.2; if RB = −10% the probability is at most 1% when the manufacturing process mean is between between 0.82 and 1.01. Whether or not these levels of risk are acceptable will depend on the ATP.

2.6.2 Proportion of out-of-specification results for manufactured lots, estimated from data

Estimation of the proportion of results from manufactured batches testing out-of-specification from data is the same as estimation of the probability of an out-of-specification result (see Equation 2.12), with the factor of $\frac{1}{(1-P)}$ applied to the variance:

$$\widehat{Prob}(\text{OOS}) = \Phi \left[\frac{\log(\text{LSL}) - \log\left(\frac{\widehat{RB}}{100} + 1\right)}{\log\left(\frac{\widehat{IP}}{100} + 1\right) \times \sqrt{\frac{n-1}{n}} \times \sqrt{\frac{1}{1-P}}} \right]$$

$$+ 1 - \Phi \left[\frac{\log(\text{USL}) - \log\left(\frac{\widehat{RB}}{100} + 1\right)}{\log\left(\frac{\widehat{IP}}{100} + 1\right) \times \sqrt{\frac{n-1}{n}} \times \sqrt{\frac{1}{1-P}}} \right].$$

Example: proportion of out-of-specification results for manufactured lots

We conduct the calculations for the data presented in Table 2.1 with 12 observations, where TRP = 1, $\widehat{RB}$ = −10.8% and $\widehat{IP}$ = 12.2% for specification limits of (0.70, 1.43). We assume that the process is centred at 1 and that $\sigma^2_{\text{Product}}$ makes up $P = 0.2$ of σ^2_{Manuf}. In this case, the estimated probability that a new single RP result will be OOS is 2.47% and is calculated as:

$$\widehat{Prob}(\text{OOS}) = \Phi \left[\frac{\log(0.70) - \log\left(\frac{-10.8}{100} + 1\right)}{\log\left(\frac{12.2}{100} + 1\right) \times \sqrt{\frac{12-1}{12}} \times \sqrt{\frac{1}{1-0.2}}} \right]$$

$$+ 1 - \Phi \left[\frac{\log(1.43) - \log\left(\frac{-10.8}{100} + 1\right)}{\log\left(\frac{12.2}{100} + 1\right) \times \sqrt{\frac{12-1}{12}} \times \sqrt{\frac{1}{1-0.2}}} \right].$$

$$= \Phi \left(\frac{-0.2424}{0.1232}\right) + 1 - \Phi \left(\frac{0.4720}{0.1232}\right)$$

$$= \Phi \left(-1.9671\right) + 1 - \Phi \left(3.8302\right)$$

$$= 0.0246 + 0.0001$$

$$= 0.0247.$$

For the upper 95% (one-sided) confidence limit:

1. Calculate:

$$K_L = \frac{[\hat{\mu}_{\text{Manuf}} - \log(\text{LSL})]}{\hat{\sigma}_{\text{Manuf}}}$$

$$= \frac{[\log(0.892) - \log(0.70)]}{\log(1.122) \times \sqrt{\frac{1}{1-0.2}}}$$

$$= 1.883$$

$$K_U = \frac{[\log(\text{USL}) - \hat{\mu}_{\text{Manuf}}]}{\hat{\sigma}_{\text{Manuf}}}$$

$$= \frac{[\log(1.43) - \log(0.892)]}{\log(1.122) \times \sqrt{\frac{1}{1-0.2}}}$$

$$= 3.667$$

Therefore, $\max(K_L, K_U) = 3.667 > \frac{(n-1)}{n} = 3.175$.

Set: $K^* = \min(K_L, K_U) = 1.883$.

5. Find:

$$\lambda : Prob\left(t_{\lambda;11} < \sqrt{12} \times 1.883\right) = 0.95$$

$$\lambda = 3.641.$$

6. Finally:

$$U = 1 - \Phi\left(\frac{3.641}{\sqrt{12}}\right)$$

$$= 14.66\%.$$

The upper (one-sided) 95% confidence limit for the probability that a new single RP result will be OOS is 14.66%. This limit is a considerably larger than the estimate of 2.47%.

2.6.3 Prediction interval for a batch from the manufacturing distribution

As in Section 2.5.4, we may be interested in forming a prediction interval for a batch from the manufacturing distribution. Again, the quantity that we are interested in is the distance of the result from its true value on the log scale:

$$d = \log(\text{RP}) - \log(\text{TRP}).$$

For a random lot from the manufacturing distribution, a $100p\%$ prediction interval for d is an interval within which $100p\%$ of distances are expected to fall. The formula is the same as in Equation (2.9) but with $\mu_{\log \text{RP}}$ replaced with:

$$\mu_{\text{Manuf}} = \mu_{\text{Product}} + \log\left(\frac{\text{RB}}{100} + 1\right),$$

TABLE 2.14

Summary of calculations for the 99% prediction interval for distance from true value: RB = -5%, IP = 10%, assuming the product variance is 20% of the total variance.

	Calculations on $\log_e$ scale	Original scale
Relative bias		-5.0%
Intermediate precision		10.0%
Distance from true value	-0.0513	0.950
Standard deviation	0.0953	
P	0.20	
$z_{0.995}$	2.5758	
99% Prediction interval	$(-0.3258, 0.2232)$	$(0.722, 1.250)$

and $\sigma^2_{\log \mathrm{RP}}$ replaced with:

$$\sigma^2_{\mathrm{Manuf}} = \frac{\left[\log\left(\frac{\mathrm{IP}}{100} + 1\right)\right]^2}{(1 - P)}.$$

Thus, for a random lot from the manufacturing distribution, a $100p\%$ prediction interval for d is given by:

$$[\mu_{\mathrm{Manuf}} - \log(\mathrm{TRP})] \pm z_{\frac{(1+p)}{2}} \times \sqrt{\sigma^2_{\mathrm{Manuf}}}$$

or equivalently, in terms of RB and IP:

$$\mu_{\mathrm{Product}} + \log\left(\frac{\mathrm{RB}}{100} + 1\right) \pm z_{\frac{(1+p)}{2}} \times \sqrt{\frac{\left[\log\left(\frac{\mathrm{IP}}{100} + 1\right)\right]^2}{1 - P}}.$$

When the manufacturing process is centred at RP = 1 the interval becomes:

$$\log\left(\frac{\mathrm{RB}}{100} + 1\right) \pm z_{\frac{(1+p)}{2}} \times \sqrt{\frac{\left[\log\left(\frac{\mathrm{IP}}{100} + 1\right)\right]^2}{1 - P}}.$$

Table 2.14 summarises the calculation of the 99% prediction interval for values of the distance of the result from the true value when the accuracy and precision are both known at RB = -5% and IP = 10% and the product variance is 20% of the total variance. The back-transformed interval represents multiples of the true value on the original scale. Thus, 99% of measured RP results are expected to fall within 72.2% and 125.0% of their true value.

2.6.3.1 Prediction interval for a batch from the manufacturing distribution, estimated from data

The prediction interval is estimated from a set of n data values as:

$$[\widehat{\mu}_{\log \mathrm{RP}} - \log(\mathrm{TRP})] \pm t_{\frac{(1+p)}{2},(n-1)} \times \sqrt{\frac{\widehat{\sigma}^2_{\log \mathrm{RP}} \times \left(1 + \frac{1}{n}\right)}{1 - P}}.$$

TABLE 2.15
Summary of calculations for 99%/95% tolerance interval for the result based on the data in
Table 2.1, assuming the product variance is 20% of the total variance.

	Calculations on $\log_e$ scale	Back-transformed
Relative bias		-10.8%
Intermediate precision		12.2%
Relative potency	-0.1143	0.892
Standard deviation	0.1151	
P	0.20	
Sample size, n	12	
$z_{0.995}$	2.5758	
$\chi^2_{0.05,11}$	4.5748	
99%/95% tolerance interval	$(-0.6493, 0.4208)$	$(0.522, 1.523)$

The 99% prediction interval for values of the distance of the result from the true value
when the accuracy and precision are based on the data presented in Table 2.1, assuming
the product variance is 20% of the total variance, is (0.588 1.32). The details of these
calculations are not shown.

2.6.4   Tolerance interval for a batch from the manufacturing distribution

The $(p,\ 1-\alpha)$ tolerance interval (TI) for lots from the manufacturing distribution is given
by:

$$
\left[\mu_{\text{Product}} + \log\left(\frac{\widehat{\text{RB}}}{100} + 1 \right) \right] \pm z_{\frac{(1+p)}{2}} \times \sqrt{ \frac{ \left[\log\left(\frac{\widehat{\text{IP}}}{100} + 1 \right) \right]^2 \times \left(1 + \frac{1}{n}\right) \times (n-1) }{ (1-P) \times \chi^2_{\alpha, n-1} } },
$$

and when the manufacturing process is centred at RP = 1:

$$
\log\left(\frac{\widehat{\text{RB}}}{100} + 1 \right) \pm z_{\frac{(1+p)}{2}} \times \sqrt{ \frac{ \left[\log\left(\frac{\widehat{\text{IP}}}{100} + 1 \right) \right]^2 \times \left(1 + \frac{1}{n}\right) \times (n-1) }{ (1-P) \times \chi^2_{\alpha, n-1} } }.
$$

Table 2.15 summarises the calculation of the 99%/95% tolerance interval for the result
based on the data in Table 2.1, assuming the product variance is 20% of the total variance.
The tolerance interval on the back-transformed (original scale) is (0.522, 1.523). This interval
does not fall within the specification limits of (0.70, 1.43). Thus at the 95% confidence
level, we can conclude that the data do not support that at least 99% of results will fall in
specification, or equivalently that at most 1% of results will fall OOS. This is consistent with

TABLE 2.16

Probability of an OOS result and 99% prediction interval for assumed values of RB $= -5\%$ and IP $= 10\%$; manufacturing distribution centred at 1 and P $= 0.2$.

	Prob(OOS)	99% Prediction interval for bias
Lots with TRP $= 1$	0.0007	(0.743, 1.214)
Manufacturing distribution	0.0021	(0.722, 1.250)

the finding above that the upper 95% confidence limit for the proportion of OOS results is greater than 1%. The conclusion would be that (assuming the accuracy and IP cannot be improved) more than one repeat of the assay method will be required for the release procedure.

In this section, we have extended the calculations of the probability of an out-of-specification, the prediction interval and the tolerance interval from a lot with known true value to the manufacturing distribution as a whole. The calculations assume given values of the performance characteristics.

2.7 Summary comparison of combinations of accuracy and precision

To aid comparison, Table 2.16 summarises the probability of an OOS result (from Tables 2.7 and 2.13) and the 99% prediction interval for the bias (from Tables 2.8 and 2.14), first for a fixed TRP of 1 and secondly for the manufacturing distribution. Here, the RB and IP are assumed to be known, at -5% and 10%, respectively and the specification limits are 0.70 and 1.43. The manufacturing distribution is assumed to account for 20% of the release assay variance. Since the manufacturing distribution incorporates additional variability the probability of an OOS result is larger and the prediction interval is wider. The 99% prediction intervals for the bias are both within the specification limits. This is consistent with the probabilities of OOS results being less than 1%. These measures can both be used to support the choice of assay format.

Table 2.17 summarises the probability of an OOS result and its upper 95% (one-sided) confidence limit, the 99% prediction interval and the 99%/95% tolerance interval for RP, first for a fixed TRP of 1 and secondly for the manufacturing distribution, as shown in Tables 2.11, 2.12, and 2.15 for comparison. Here, the RB and IP are estimated from the data presented in Table 2.1; their estimated values are -10.8% and 12.2%, respectively. Again, the specification limits are 0.70 and 1.43 and the manufacturing distribution is assumed to account for 20% of the release assay variance. The 99% prediction intervals for the bias both breach the specification limits at the low end, consistent with the probability of an OOS result being higher than 1%. The 99%/95% tolerance intervals breach the specification limits at both ends. The tolerance intervals can be interpreted as prediction intervals incorporating

TABLE 2.17

Probability of an OOS result, 99% prediction interval and 99%/95% tolerance interval: RB and IP estimated from data at -10.8%, 12.2%; $P = 0.2$.

	Prob(OOS)	Upper 95% CL	99% Prediction interval for bias	99%/95% Tolerance interval for RP
For lots with TRP = 1	0.0139	0.1135	(0.615, 1.294)	(0.553, 1.439)
For manufacturing distribution	0.0247	0.1466	(0.588, 1.352)	(0.522, 1.523)

the uncertainty in the estimate of IP. They are consistent with the fact that the upper confidence limit for the probability of an OOS result is greater than 1%. These measures can be used in validation studies (see Chapter 8).

2.8 Range of application of the bioassay

It is important to define the set of true potency or concentration values for which the bioassay is capable of performing adequately; that is, the range over which the result is both accurate and precise. For relative potency assays this is called the bioassay range and is expressed in terms of relative potency values. For absolute potency or concentration, these are the limits of quantitation.

2.8.1 Relative potency bioassays: range

The range of application of a relative potency bioassay is the range of true potencies for which the bioassay is both accurate and precise. Alternatively, this could be the range of true potencies for which a tolerance interval is acceptably narrow.

If a result for a batch falls outside the range of application, then the batch can be re-tested, either at a higher or a lower starting concentration in order to bring the result within the range. Once a result is achieved which does fall within the range of application, it can be adjusted back to correspond to the original concentration by multiplication.

2.8.2 Concentration bioassays: LOD, LLOQ, ULOQ

Where a bioassay measures an absolute potency or concentration, limits of quantitation and detection are required: the limit of detection (LOD), lower limit of quantitation (LLOQ) and the upper limit of quantitation (ULOQ).

The limit of detection is the lowest amount of analyte in a sample that can be detected, but not necessarily quantitated, under the stated experimental conditions [62]. The limit of quantitation is the lowest amount of analyte in a sample that can be determined with acceptable accuracy and precision under the stated experimental conditions [62]. These limits are driven by the ability of the assay to distinguish between blank samples (with a concentration of zero) and samples with a non-zero concentration of the analyte. The ULOQ is set based on limits on the accuracy and precision of the estimate.

The following methodology for estimating the LOD and LLOQ is recommended by USP [62]. We assume that the responses for a blank (concentration of 0) are normally distributed with mean μ_B and variance σ^2. The critical response, R_C, that has a low chance, α, of being exceeded when the true concentration is zero is found as the upper $(1-\alpha)$ quantile of the normal distribution for the blank response:

$$R_C = \mu_B + z_{1-\alpha} \times \sigma,$$

where z_p is the $100p$ percentile of the standard normal distribution.

Figure 2.16 illustrates this methodology for two samples with low concentrations. Here R_C was chosen by fixing the value of $\alpha = 0.05$ (shaded in light grey). For a sample with a very low concentration (shown in (a)), with mean response R_{D_1}, the distributions of the blank and sample responses have a considerable overlap. Therefore, it is difficult to establish a boundary that can differentiate between the two distributions. However, for a sample with a higher concentration (shown in (b)), with mean response R_{D_2}, the overlapping region is reduced and the two distributions can more easily be separated.

The LOD is a concentration for which the distribution of responses has minimal overlap with the distribution of blank responses. We define the LOD as the concentration for which the response has a low chance, β, of falling below R_C. Referring to Figure 2.16, the distance between R_C and R_{D_1} is $(z_{1-\beta} \cdot \sigma)$. Fixing β to the required value allows the mean response at the LOD, R_{LOD}, to be calculated as:

$$R_{\mathrm{LOD}} = R_C + z_{1-\beta} \times \sigma$$
$$= \mu_B + (z_{1-\alpha} + z_{1-\beta}) \times \sigma.$$

Assuming the relationship between the concentration and mean response (R) is a straight line:

$$R = \mu_B + m \times \text{concentration},$$

the concentration with mean response R_D, i.e., the LOD, is:

$$\mathrm{LOD} = \frac{R_{\mathrm{LOD}} - \mu_B}{m}$$
$$= (z_{1-\alpha} + z_{1-\beta}) \times \frac{\sigma}{m}.$$

Figure 2.17 shows the linear relationship between concentration and response with the normal curves overlaid for the blank and the LOD.

To estimate the LOD, estimates of σ and m are required. Let $(x_1, \ldots, x_n)$ represent the concentrations for a series of n samples. A straight line can be fitted to the responses versus

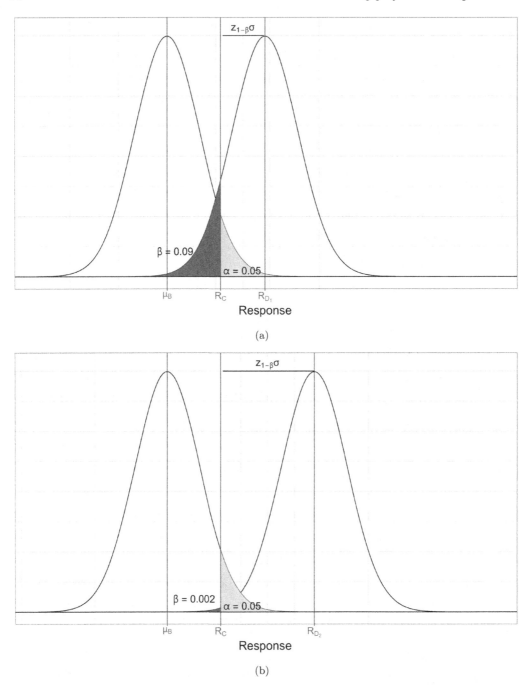

(a)

(b)

FIGURE 2.16

Distributions of response values for a blank (on the left) and a low-concentration sample on the right (for both plots). R_C was chosen by fixing the value of $\alpha = 0.05$ (shaded in light grey). This results in $\beta = 0.09$ for the lower-concentration sample in (a) and $\beta = 0.002$ for the higher-concentration sample in (b). The LOD is chosen by ensuring there is sufficient separation between the distributions of the blank and low concentration responses, i.e., fixing β at a selected level.

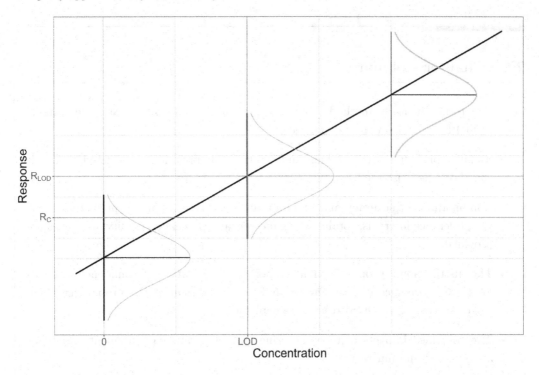

FIGURE 2.17

Relationship between the mean response and the concentration. For a given concentration, the response is normally distributed, shown by the bell curves. R_C is the 95th percentile of the distribution of responses when the concentration is 0. At the LOD, the 100β percentile of the response distribution matches R_C.

the concentrations and the uncertainties in the estimates can be handled using prediction limits. The LOD is estimated as:

$$\text{LOD} = (t_{1-\alpha,n-2} + t_{1-\beta,n-2}) \times \frac{\hat{\sigma}}{\hat{m}} \sqrt{1 + \frac{1}{n} + \frac{\bar{x}^2}{\sum_{i=1}^{n}(x_i - \bar{x})^2}}, \qquad (2.17)$$

where $\bar{x}$ is the mean concentration over $(x_1, \ldots, x_n)$, $\hat{m}$ is the estimate of the slope of the regression line, $\hat{\sigma}$ is the root-mean-squared error of the regression line (see Appendix A), and $t_{p,n-2}$ is the $100p$ percentile of the t distribution with $n-2$ degrees of freedom.

The LLOQ is generally set above the LOD. Formula 2.17 can be used, replacing $(t_{1-\alpha,n-2} + t_{1-\beta,n-2})$ with a larger value, often 10.

Other approaches to setting the LOD and LLOQ are available. For example, ICH Q2R2 [27] presents the following formulae:

$$\text{LOD} = 3.3 \frac{\hat{\sigma}}{\hat{m}},$$

$$\text{LLOQ} = 10 \frac{\hat{\sigma}}{\hat{m}}.$$

Note that if α and β are both set at 0.05, $z_{1-\alpha} + z_{1-\beta} = 3.3$. These formulae are simplifications of the former approach with no account taken of the uncertainty in the estimates.

2.9   Chapter summary

- The terms 'bioassay method', 'bioassay procedure' and 'analytical target profile' (ATP) for a bioassay are introduced.

- 'Relative bias', 'proportional bias' and 'intermediate precision' are defined, and their calculation is illustrated.

- Combinations of accuracy and precision are presented, including prediction intervals, tolerance intervals, total analytical error, and process capability; examples are provided.

- The distributional properties that support the calculation of confidence intervals for the performance characteristics, and the calculation of the probability of an out-of-specification (OOS) result, are explained.

- The combined impacts of manufacturing variability and measurement variability are explored and illustrated.

3

Bioassay design

The development of a bioassay involves several stages. First, a biological system must be found which is capable of exhibiting a relationship between the dose of the test sample and its measured response. Second, the various factors which control the system, such as cell passage number, reagents, incubation conditions, age of animals or duration of assay, must be examined to find the optimal combination. Acceptable levels of variation in the factors (e.g. incubation time 10 minutes $\pm$ 2 minutes) must be determined and it is critical to ensure that the assay is robust to changes in the factor levels within the allowed limits. The levels of these factors should be chosen in an effective and efficient way, making use of design of experiments methodology. Making use of factorial experimental designs is preferable to testing one factor at a time. These stages are not covered in this book: the first involves scientific, biological considerations; to do justice to the second would require a dedicated textbook, and there are many such books available [11, 23, 40].

We start from the point where the biological system is designed and the shape of the dose-response relationship is established, at least in terms of a straight line versus a full S-shaped curve. The details of the design of the bioassay method (see Chapter 2) remain to be optimised. In this chapter, we consider the layout of samples and doses in wells on plates (or animals in cages, in blocks), and the number of plates to be included in the bioassay method, to produce a measurement of concentration or relative potency. We also consider the design of the release procedure which is driven by the analytical target profile (ATP – see Chapter 2) for the bioassay.

The design of the plate layout is important to both the accuracy and precision of the method, while the replication strategy for the assay method further contributes to precision. The discussion applies equally to relative potency assays and interpolation-type assays, where a concentration is estimated. The choice of dose groups for the test sample and reference standard is considered in Chapter 5.

3.1 Allocation of treatments to experimental units

The design of the bioassay is fundamental to the accuracy of the method. Bias can be introduced or exacerbated by an underlying response pattern on a plate, including edge effects, differences between plates, or cage positioning in the case of animal assays. We

DOI: 10.1201/9781003449195-3

	1	2	3	4	5	6	7	8	9	10	11	12
A												
B												
C												
D												
E												
F												
G												
H												

FIGURE 3.1

A blank 96-well plate consisting of 8 rows and 12 columns. The rows are indexed with letters
A through H and the columns are indexed by numbers 1 through 12.

	1	2	3	4	5	6	7	8	9	10	11	12
A	$T1_6$	$T3_4$	$T2_{10}$	$T1_{10}$	$T1_1$	S_4	$T1_{12}$	$T2_8$	$T1_3$	$T3_6$	S_7	$T3_4$
B	S_6	$T2_2$	$T3_{12}$	$T2_1$	$T1_2$	$T3_9$	$T2_3$	$T2_3$	S_{11}	S_8	S_{11}	S_9
C	$T3_{11}$	S_1	S_1	$T3_8$	$T3_1$	$T3_2$	$T3_2$	S_2	$T3_{10}$	$T2_6$	S_{10}	$T2_7$
D	$T1_4$	$T2_{11}$	$T1_{11}$	$T1_3$	$T3_{12}$	$T3_6$	$T3_3$	$T3_5$	$T3_{10}$	S_3	$T1_8$	$T2_6$
E	S_8	S_{12}	$T2_{11}$	$T1_{10}$	$T1_1$	S_4	S_9	$T1_7$	$T1_7$	S_5	$T3_1$	S_7
F	$T2_2$	$T3_8$	$T3_7$	$T1_{11}$	$T2_{12}$	$T2_{12}$	$T3_7$	$T1_{12}$	$T1_5$	S_2	$T3_9$	$T2_{10}$
G	$T2_9$	$T2_4$	$T2_5$	$T1_9$	$T1_6$	$T1_9$	$T1_5$	S_3	$T2_5$	$T1_4$	S_6	$T2_9$
H	$T2_1$	$T3_5$	$T2_4$	$T3_{11}$	$T3_3$	$T2_7$	$T1_2$	S_5	S_{10}	$T2_8$	S_{12}	$T1_8$

FIGURE 3.2

A 96-well plate where all combinations of sample and dose are allocated to a well completely
at random. S refers to the reference standard, T1 – T3 refer to the three test samples and
the shading and subscript refer to the dilution.

focus our discussion on the microtitre plate case, where several plates might be used to
provide a single measurement. Similar arguments apply to blocks of cages in an animal
assay.

3.1.1   Plate layout: 96-well plate

We will focus our discussion on a 96-well plate. However, the same principles will apply to
other plate structures. The wells of a 96-well plate can be indexed by 8 rows (A through H)
and 12 columns (1 through 12) as shown in Figure 3.1.

The ideal design from a statistical point of view might have all the combinations of
sample and dose allocated completely at random to the 96 wells on the plate. Figure 3.2
shows an example with a reference standard, S, and three test samples, T1, T2, T3, each
tested at 12 doses with 2 replicates each. The doses are indicated by the subscripts 1 to
12 as well as by the shading intensity. This avoids creating bias where there are systematic

	1	2	3	4	5	6	7	8	9	10	11	12
A	S_1	S_2	S_3	S_4	S_5	S_6	S_7	S_8	S_9	S_{10}	S_{11}	S_{12}
B	S_1	S_2	S_3	S_4	S_5	S_6	S_7	S_8	S_9	S_{10}	S_{11}	S_{12}
C	$T1_1$	$T1_2$	$T1_3$	$T1_4$	$T1_5$	$T1_6$	$T1_7$	$T1_8$	$T1_9$	$T1_{10}$	$T1_{11}$	$T1_{12}$
D	$T1_1$	$T1_2$	$T1_3$	$T1_4$	$T1_5$	$T1_6$	$T1_7$	$T1_8$	$T1_9$	$T1_{10}$	$T1_{11}$	$T1_{12}$
E	$T2_1$	$T2_2$	$T2_3$	$T2_4$	$T2_5$	$T2_6$	$T2_7$	$T2_8$	$T2_9$	$T2_{10}$	$T2_{11}$	$T2_{12}$
F	$T2_1$	$T2_2$	$T2_3$	$T2_4$	$T2_5$	$T2_6$	$T2_7$	$T2_8$	$T2_9$	$T2_{10}$	$T2_{11}$	$T2_{12}$
G	$T3_1$	$T3_2$	$T3_3$	$T3_4$	$T3_5$	$T3_6$	$T3_7$	$T3_8$	$T3_9$	$T3_{10}$	$T3_{11}$	$T3_{12}$
H	$T3_1$	$T3_2$	$T3_3$	$T3_4$	$T3_5$	$T3_6$	$T3_7$	$T3_8$	$T3_9$	$T3_{10}$	$T3_{11}$	$T3_{12}$

FIGURE 3.3

A 96-well plate where each row consists of a complete dilution series for a sample, where the sample replicates are placed in adjacent rows. Such a design has the potential to introduce bias if there are any row to row difference. S refers to the reference standard, T1 – T3 refer to the three test samples and the shading and subscript refer to the dilution.

underlying differences across the plate. However, this design is rarely feasible, and it would not make the best use of the plate when systematic differences are present.

3.1.1.1 Simple plate layouts

Plate layouts are typically designed for simple and error-free execution of the bioassay, particularly when the dosing is manual. An example is shown in Figure 3.3 for a similar bioassay with a reference standard and 3 test samples as above. In this example, the doses have been placed on the plate in increasing dilution order from left to right for all samples and the sample replicates have been placed in adjacent rows.

An extremely simple layout like this has the potential to introduce bias in the estimates of potency if there are any intrinsic variations from row to row. A simple improvement would be to create two 'blocks' in the plate consisting of the top 4 rows and the bottom 4 rows with each sample appearing once in each block as shown in Figure 3.4. For each test sample, this provides two comparisons with the reference standard – one within each block. If there are systematic difference between the rows, the combined result will be more accurate and more precise than the result from the plate layout in Figure 3.3, where the replicates for each sample are placed in adjacent rows. (Note that to take full advantage of this design, the statistical analysis must account for the blocked structure.)

3.1.1.2 Uniformity experiment

To mitigate the risk of bias, it is important to understand any intrinsic variation occurring across the plate. A useful starting point is to run a 'uniformity experiment' with all wells in all plates containing the same dose of a single test sample, for example, the reference standard. Ideally, the responses should show random variation, with no patterns in evidence. Patterns can be visually assessed via a simple 'heat map'. This can be constructed in Excel [38] and shades the wells according to the rank of their response value.

	1	2	3	4	5	6	7	8	9	10	11	12
A	S_1	S_2	S_3	S_4	S_5	S_6	S_7	S_8	S_9	S_{10}	S_{11}	S_{12}
B	$T1_1$	$T1_2$	$T1_3$	$T1_4$	$T1_5$	$T1_6$	$T1_7$	$T1_8$	$T1_9$	$T1_{10}$	$T1_{11}$	$T1_{12}$
C	$T2_1$	$T2_2$	$T2_3$	$T2_4$	$T2_5$	$T2_6$	$T2_7$	$T2_8$	$T2_9$	$T2_{10}$	$T2_{11}$	$T2_{12}$
D	$T3_1$	$T3_2$	$T3_3$	$T3_4$	$T3_5$	$T3_6$	$T3_7$	$T3_8$	$T3_9$	$T3_{10}$	$T3_{11}$	$T3_{12}$
E	S_1	S_2	S_3	S_4	S_5	S_6	S_7	S_8	S_9	S_{10}	S_{11}	S_{12}
F	$T1_1$	$T1_2$	$T1_3$	$T1_4$	$T1_5$	$T1_6$	$T1_7$	$T1_8$	$T1_9$	$T1_{10}$	$T1_{11}$	$T1_{12}$
G	$T2_1$	$T2_2$	$T2_3$	$T2_4$	$T2_5$	$T2_6$	$T2_7$	$T2_8$	$T2_9$	$T2_{10}$	$T2_{11}$	$T2_{12}$
H	$T3_1$	$T3_2$	$T3_3$	$T3_4$	$T3_5$	$T3_6$	$T3_7$	$T3_8$	$T3_9$	$T3_{10}$	$T3_{11}$	$T3_{12}$

FIGURE 3.4

A 96-well plate where each the samples are arranged in a block design. Each sample appears once in the top block (first four rows) and once in the bottom block (last four rows). Such a design has the potential to mitigate bias due to row differences. S refers to the reference standard, T1 – T3 refer to the three test samples and the shading and subscript refer to the dilution.

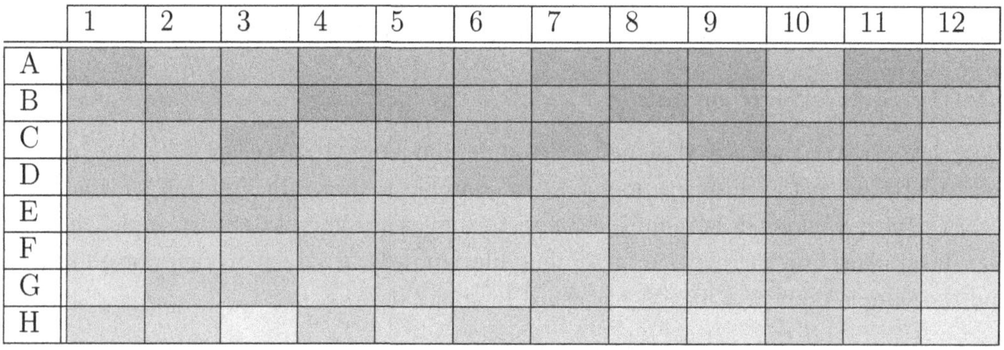

FIGURE 3.5

A heat map for the response values of a uniformity experiment. The shading intensity represents the value of the response, the darker the shade, the higher the response. In general, higher values of the response can be found at the top of the plate.

Figure 3.5 shows a heat map for a sample data set; the darkest well (and largest response) is A11 and the palest (and smallest response) is H3. In general, there is a trend towards lower values at the bottom of the plate. If a pattern such as this is seen, then attention needs to be given to the conduct of the bioassay – there could be operations that are carried out in a particular order which is causing the trend across the plate. If it is not possible to remove the background pattern, then the plate layout should be designed to ensure any potential bias is minimised.

3.1.1.3 Layouts informed by uniformity experiment

If a trend in the response value across the plate is found, as seen in Figure 3.5, with higher responses at the top and lower responses at the bottom, then it is possible to arrange the

	1	2	3	4	5	6	7	8	9	10	11	12
A	S_1	S_2	S_3	S_4	S_5	S_6	S_7	S_8	S_9	S_{10}	S_{11}	S_{12}
B	$T1_1$	$T1_2$	$T1_3$	$T1_4$	$T1_5$	$T1_6$	$T1_7$	$T1_8$	$T1_9$	$T1_{10}$	$T1_{11}$	$T1_{12}$
C	$T2_1$	$T2_2$	$T2_3$	$T2_4$	$T2_5$	$T2_6$	$T2_7$	$T2_8$	$T2_9$	$T2_{10}$	$T2_{11}$	$T2_{12}$
D	$T3_1$	$T3_2$	$T3_3$	$T3_4$	$T3_5$	$T3_6$	$T3_7$	$T3_8$	$T3_9$	$T3_{10}$	$T3_{11}$	$T3_{12}$
E	$T3_1$	$T3_2$	$T3_3$	$T3_4$	$T3_5$	$T3_6$	$T3_7$	$T3_8$	$T3_9$	$T3_{10}$	$T3_{11}$	$T3_{12}$
F	$T2_1$	$T2_2$	$T2_3$	$T2_4$	$T2_5$	$T2_6$	$T2_7$	$T2_8$	$T2_9$	$T2_{10}$	$T2_{11}$	$T2_{12}$
G	$T1_1$	$T1_2$	$T1_3$	$T1_4$	$T1_5$	$T1_6$	$T1_7$	$T1_8$	$T1_9$	$T1_{10}$	$T1_{11}$	$T1_{12}$
H	S_1	S_2	S_3	S_4	S_5	S_6	S_7	S_8	S_9	S_{10}	S_{11}	S_{12}

FIGURE 3.6

A 96-well plate where the replicate rows for each sample have been distributed across the plate so that, on average, each sample is in approximately the centre row. This design can help to mitigate bias resulting from a trend down the plate (from top to bottom). S refers to the reference standard, T1 – T3 refer to the three test samples and the shading and subscript refer to the dilution.

	1	2	3	4	5	6	7	8	9	10	11	12
A	S_1	S_2	S_3	S_4	S_5	S_6	S_7	S_8	S_9	S_{10}	S_{11}	S_{12}
B	$T1_1$	$T1_2$	$T1_3$	$T1_4$	$T1_5$	$T1_6$	$T1_7$	$T1_8$	$T1_9$	$T1_{10}$	$T1_{11}$	$T1_{12}$
C	$T2_1$	$T2_2$	$T2_3$	$T2_4$	$T2_5$	$T2_6$	$T2_7$	$T2_8$	$T2_9$	$T2_{10}$	$T2_{11}$	$T2_{12}$
D	$T3_1$	$T3_2$	$T3_3$	$T3_4$	$T3_5$	$T3_6$	$T3_7$	$T3_8$	$T3_9$	$T3_{10}$	$T3_{11}$	$T3_{12}$
E	$T3_{12}$	$T3_{11}$	$T3_{10}$	$T3_9$	$T3_8$	$T3_7$	$T3_6$	$T3_5$	$T3_4$	$T3_3$	$T3_2$	$T3_1$
F	$T2_{12}$	$T2_{11}$	$T2_{10}$	$T2_9$	$T2_8$	$T2_7$	$T2_6$	$T2_5$	$T2_4$	$T2_3$	$T2_2$	$T2_1$
G	$T1_{12}$	$T1_{11}$	$T1_{10}$	$T1_9$	$T1_8$	$T1_7$	$T1_6$	$T1_5$	$T1_4$	$T1_3$	$T1_2$	$T1_1$
H	S_{12}	S_{11}	S_{10}	S_9	S_8	S_7	S_6	S_5	S_4	S_3	S_2	S_1

FIGURE 3.7

A 96-well plate where the replicate rows for each sample have been distributed across the plate so that, on average, each sample is in approximately the centre row and the dilution series for the top four rows goes left to right and is reversed for the bottom four rows. This design can help to mitigate bias resulting from trends going both down (from top to bottom) and across (from left to right) the plate. S refers to the reference standard, T1 – T3 refer to the three test samples and the shading and subscript refer to the dilution.

samples in a way to mitigate the impact of this variation. Figure 3.6, gives an example layout that would help to remove the bias caused by a background trend down the columns. Here, the two replicates for each sample have been distributed across the plate so that, on average, each sample is in approximately in the centre of the plate. If a pattern over the columns (from left to right) is also evident, reversing the order of the dilutions for the two replicates can also help to reduce bias as shown in Figure 3.7.

More sophisticated designs are generally too complex for manual dosing, but may be possible with a robotic system. Examples include so-called split-plot and strip-plot designs

TABLE 3.1

Example of a distribution of the samples across a four plate assay to mitigate impact of background patterns.

Row	Plate 1	Plate 2	Plate 3	Plate 4
A	S	T2	T1	T3
B	T1	T3	T2	S
C	T2	S	T3	T1
D	T3	T1	S	T2
E	T3	T1	S	T2
F	T2	S	T3	T1
G	T1	T3	T2	S
H	S	T2	T1	T3

[31, 60]. In a split-plot design, the samples are allocated to rows at random; within each row, the doses are allocated at random. In a strip-plot design, the samples are allocated to rows at random; the starting dose for the dilution series is allocated to a column at random (and is the same for each row).

3.1.1.4 Several plates

In a bioassay where several replicate plates are needed in order to accommodate the required number of test samples, or the required number of replicates per test sample, there is an opportunity to vary the positions of the samples within each plate. This can further mitigate any bias caused by background patterns. Table 3.1 shows an example of how the samples could be distributed across a four plate assay.

3.2 Precision and replication

The concept of precision is key in the design of a bioassay. Precision can be improved by replication. The most efficient use of material is made when replication is conducted for conditions with the most variation: there is no point in replication if the results are expected to be close to each other. In particular, it is rarely of benefit to include technical replicates, or pseudo-replicates. All replicates should be independently prepared (see Chapter 5).

3.2.1 Variance components

As in Chapter 2, we present the concepts in terms of a measurement of relative potency, RP. However, the same approach can be applied to concentrations.

A single execution, or run, of the bioassay method results in a measurement. The method run can include a single plate, or a set of plates executed under similar conditions. The plates can include replicates of the test sample. We assume the distribution of the measurements across multiple executions of the method (runs) is log normal:

$$\log(\text{RP}) \sim N\left(\mu_{\log \text{RP}}, \sigma^2_{\log \text{RP}}\right).$$

The variance of this distribution, $\sigma^2_{\log \text{RP}}$, is made up of contributions from several potential sources:

- **Within runs**

  - Between plates within an execution of the method, the variability comes from plate differences and any other contributions such as time of day, temperature and so on.

  - Within a plate, the variability comes from the replicates at each dose (sometimes called pure error) and from the fit of the data to the dose-response relationship.

- **Between runs**

  - The between-run variance component can itself be made up of levels of variability. For example, if the whole method can be run several times on the same day by the same operator, the variance is likely to be lower than if the method is run by different operators on different days. We will refer to a set of runs conducted on the same day by the same operator as a 'session'.

Recall from Chapter 2 that the IP is the value of GCV for a single execution of the method. Assessment of this measure of precision needs to allow for the maximum amount of likely variability amongst runs to avoid an overly optimistic assessment of the bioassay performance. Suppose this maximum variability occurs amongst runs conducted in different sessions. Then, the relevant variance components are:

$$\sigma^2_{\log \text{RP}} = \sigma^2_{\text{Between sessions}} + \sigma^2_{\text{Within session}}. \tag{3.1}$$

The intermediate precision is then given by:

$$\text{IP} = 100 \times \left[\text{antilog}\left(\sqrt{\sigma^2_{\text{Between sessions}} + \sigma^2_{\text{Within session}}}\right) - 1\right] \%. \tag{3.2}$$

The within session variance component, $\sigma^2_{\text{Within session}}$, can be thought of as akin to repeatability as defined for small molecule assays [27].

3.3 Quality control sample

As well as the reference standard, it is valuable to include a quality control (QC), or assay control, sample on each plate. This sample is a batch of material with a 'known' (that is, well-estimated) potency relative to the reference standard. Such a sample is valuable in demonstrating whether or not the plate has performed as expected because its result is a fixed value (see Chapter 7).

The QC sample is often a replicate of the reference standard itself, in which case its measured potency is expected to be 1. Another approach is to use a candidate reference batch, in which case the transition to the new reference batch when the original expires is simplified because the relative potency of the new reference standard is already well-estimated (see Chapter 10).

3.4   Design of bioassay release procedure

If the GCV for a single execution of the method (i.e., the intermediate precision, see Chapter 2) is unacceptably high, replication of the method is required to provide the necessary performance for the release procedure. In other words, the assay release procedure may involve several runs of the method; these may be within a single session or spread over several sessions. If the method is independently repeated n times and the results are averaged (on the log scale), then the variance of the estimate $\hat{\mu}_{\log \text{RP}}$ will be:

$$Var\left(\hat{\mu}_{\log \text{RP}}\right) = \frac{\sigma^2_{\log \text{RP}}}{n}.$$

In terms of the intermediate precision:

$$Var\left(\hat{\mu}_{\log \text{RP}}\right) = \frac{\left[\log\left(\frac{\text{IP}}{100} + 1\right)\right]^2}{n}. \tag{3.3}$$

If the repeated measurements are not independent, for example, if they come from multiple runs within the same session, the components of variance should be accounted for.

Suppose the relative bias is 10%, the intermediate precision is 20%, and the specification limits for release are 0.70 and 1.43. Then Figure 3.8 shows the effect of repeating the bioassay method on the distribution of the final estimate, $\hat{\mu}_{\log \text{RP}}$. Overall, it can be seen that a smaller proportion of the distribution falls outside of the specification limits as the number of replicates is increased. The probabilities of a result being outside the specification limits are 8.2%, 2.1%, and 0.6% for 1, 2, and 3 replicates, respectively. These probabilities can be calculated using Equation (2.8), substituting in the variance calculated as per Equation (3.3). To achieve a probability of an out-of-specification result of at most 1%, for a test sample with TRP = 1, at least 3 repeats of the bioassay method are required in this case.

3.4.1   Reportable value for release

We define the release procedure format as n_F sessions, each including k_F runs of the method. We define the combined measurements from the $n_F \times k_F$ runs of the method as the reportable value (RV), sometimes called the 'reportable result'. Its variance on the log scale, the format variability, is given by the following combination of the variance components:

$$\sigma^2_{\log \text{RV}} = \frac{\sigma^2_{\text{Between sessions}}}{n_F} + \frac{\sigma^2_{\text{Within session}}}{k_F \times n_F}. \tag{3.4}$$

Suppose the procedure for lot release is for each of three operators to conduct the method twice within a day. Then $k_F = 2$ and $n_F = 3$.

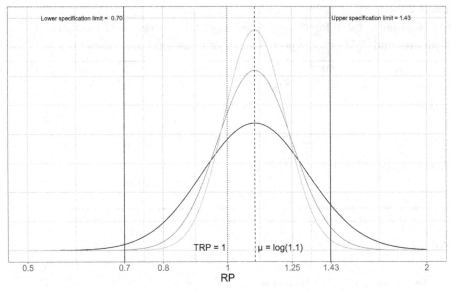

FIGURE 3.8

Distribution for final assay release results combined over 1, 2, or 3 replicate runs, where $\mu_{\log \mathrm{RP}} = \log(1.1)$, IP = 20%, and TRP = 1. The probabilities of a result being outside the specification limits are 8.2%, 2.1%, and 0.6% for 1, 2, and 3 replicates, respectively.

3.4.2   Estimation of the variance of the reportable value

Estimates of the variance components allow us to compare replication strategies, in the light of practical concerns such as how many runs of the method can feasibly be executed by one operator in a day. The optimal values for the assay format n_F and k_F can then be identified.

Suppose we have a balanced data set (the same number of results per session) including k runs of the method in each of n repeat sessions, for a single test sample. Let the relative potency estimates for each method run be denoted by RP_{ij}, where $i = 1, \ldots, n$ indexes the session and $j = 1, \ldots, k$ indexes the method run within the session. Then let $\log(\mathrm{RP}_{ij})$ be the log-transformed relative potency. We can then define $\widehat{\log \mathrm{RP}}_{i\cdot}$ to be the mean RP on the log scale for session i and $\widehat{\log \mathrm{RP}}_{\cdot\cdot}$ as the overall mean across all runs. Then, the estimation of the variance components is straightforward using analysis of variance (ANOVA) methods [8]. Table 3.2 shows the ANOVA table for the balanced case. The variance estimates can then be found as:

- Within session variance:

$$\hat{\sigma}^2_{\text{Within session}} = \mathrm{MSS}_{\text{Within}}, \tag{3.5}$$

- Between session variance:

$$\hat{\sigma}^2_{\text{Between sessions}} = \frac{(\mathrm{MSS}_{\text{Between}} - \mathrm{MSS}_{\text{Within}})}{k}, \tag{3.6}$$

TABLE 3.2

Analysis of variance table for the balanced case. DF = degrees of freedom, MSS = mean sum of squares and $\mathbb{E}[\text{MSS}]$ = expected value of mean sum of squares.

Source	DF	MSS	$\mathbb{E}[\text{MSS}]$
Between sessions	$n-1$	$\text{MSS}_{\text{Between}} =$ $k \times \dfrac{\sum_{i=1}^{n}\left(\widehat{\log \text{RP}}_{i.} - \widehat{\log \text{RP}}_{..}\right)^2}{(n-1)}$	$k\sigma^2_{\text{Between session}}$ $+\sigma^2_{\text{Within session}}$
Within session	$n(k-1)$	$\text{MSS}_{\text{Within}} =$ $\dfrac{\sum_{i=1}^{n}\sum_{j=1}^{k}\left(\log \text{RP}_{ij} - \widehat{\log \text{RP}}_{i.}\right)^2}{n(k-1)}$	$\sigma^2_{\text{Within session}}$

- Total variance:

$$\hat{\sigma}^2_{\text{Between sessions}} + \hat{\sigma}^2_{\text{Within session}} = MSS_{\text{Between}}\frac{1}{k} + \text{MSS}_{\text{Within}}\left(1 - \frac{1}{k}\right). \qquad (3.7)$$

When either the data set is unbalanced (different numbers of runs per session), or the formula gives a negative value for the estimate of the between session variance, the calculations are less straightforward. In either case, restricted maximum likelihood (REML) can be used. Standard statistical packages, including R [49], SAS [50], SPSS [26], Minitab [39] can provide the estimates of the variance components.

The estimate of intermediate precision is given by substituting the estimates of the variance components into the formula for IP (Equation (3.2)):

$$\widehat{\text{IP}} = 100 \times \left[\text{antilog}\left(\sqrt{\hat{\sigma}^2_{\text{Between sessions}} + \hat{\sigma}^2_{\text{Within session}}}\right) - 1\right]\%.$$

The estimate for the variance of the reportable value on the log scale can be found by substituting the estimates for the between and within session variance into Equation (3.4):

$$\hat{\sigma}^2_{\log \text{RV}} = \frac{\hat{\sigma}^2_{\text{Between sessions}}}{n_F} + \frac{\hat{\sigma}^2_{\text{Within session}}}{k_F \times n_F}.$$

Note that the estimated variance for the reportable value depends on both the intended number of sessions, n_F, and runs per session, k_F, as well as the variance components estimated from a data set with n sessions and k runs per session. It is not necessary for these to be equal.

To obtain a confidence interval for the variance of the reportable value on the log scale we follow the methodology in [8]. The variance of the reportable value on the log scale can be written as:

$$\sigma^2_{\log \text{RV}} = a \times \sigma^2_{\text{Between sessions}} + b \times \sigma^2_{\text{Within session}},$$

where $a = \frac{1}{n_F}$ and $b = \frac{1}{k_F \times n_F}$. The estimate of $\sigma^2_{\log \text{RV}}$ is given by:

$$\hat{\sigma}^2_{\log \text{RV}} = a \times \frac{(\text{MSS}_{\text{Between}} - \text{MSS}_{\text{Within}})}{k} + b \times \text{MSS}_{\text{Within}}$$

$$= \frac{a}{k} \times \text{MSS}_{\text{Between}} + \left(b - \frac{a}{k}\right) \times \text{MSS}_{\text{Within}}.$$

The upper (one-sided) 95% confidence limit can then be found as:

$$\hat{\sigma}^2_{\log \text{RV}} + \sqrt{H_1^2 \left(\frac{a}{k}\right)^2 \text{MSS}^2_{\text{Between}} + H_2^2 \left(b - \frac{a}{k}\right)^2 \text{MSS}^2_{\text{Within}}},$$

where:

$$H_1 = \frac{n-1}{\chi^2_{0.05,(n-1)}} - 1$$

$$H_2 = \frac{n(k-1)}{\chi^2_{0.05,n(k-1)}} - 1,$$

and $\chi^2_{p,df}$ denotes the $100p$ percentile of chi-squared distribution with df degrees of freedom. The numerators of H_1 and H_2 as well as the degrees of freedom for the χ^2 value are specific to this example. In general, these are the degrees of freedom for the $\text{MSS}_{\text{Between}}$ and $\text{MSS}_{\text{Within}}$, respectively.

As a simpler alternative, commonly available in statistical packages, calculations based on the chi-squared distribution and the Satterthwaite degrees of freedom can be used but may not offer accurate coverage in all circumstances [8]. Using the Satterthwaite approach [51], the upper (one-sided) 95% confidence limit for the variance is given by:

$$\hat{\sigma}^2_{\log \text{RV}} \times \frac{df}{\chi^2_{0.05,df}},$$

where df is the Satterthwaite degrees of freedom:

$$df = \frac{\hat{\sigma}^4_{\log \text{RV}}}{\frac{\left[\frac{a}{k} \times \text{MSS}_{\text{Between}}\right]^2}{n-1} + \frac{\left[\left(b - \frac{a}{k}\right) \times \text{MSS}_{\text{Within}}\right]^2}{n \cdot (k-1)}}.$$

The denominators in the denominator ($n-1$ and $n \cdot (k-1)$) are the degrees of freedom for the $\text{MSS}_{\text{Between}}$ and $\text{MSS}_{\text{Within}}$, respectively. In general, the degrees of freedom for these calculations will be fractional instead of an integer value. Some software packages, including Excel [38], take the floor (i.e., ignore any digits after the decimal place).

Example: estimating the variance of the reportable value

Suppose the data set presented in Chapter 2 Table 2.1 was obtained by conducting $n = 4$ sessions each consisting of $k = 3$ method runs. These data are presented in Table 3.3 and plotted in Figure 3.9.

Table 3.4 provides the analysis of variance table for the example data. Referring to Equations (3.5)–(3.7), the variance components are estimated as follows. Estimate of the within session variance:

$$\hat{\sigma}^2_{\text{Within session}} = 0.00201.$$

TABLE 3.3

RP results for twelve method runs broken down as $n = 4$ sessions, each with $k = 3$ method runs.

Session	Method run	Result RP	Result log RP	Run mean
	1	0.78	−0.2485	
1	2	0.80	−0.2231	−0.2531
	3	0.75	−0.2877	
	1	0.91	−0.0943	
2	2	0.92	−0.0834	−0.0728
	3	0.96	−0.0408	
	1	1.06	0.0583	
3	2	0.95	−0.0513	0.0280
	3	1.08	0.0770	
	1	0.83	−0.1863	
4	2	0.84	−0.1744	−0.1591
	3	0.89	−0.1165	

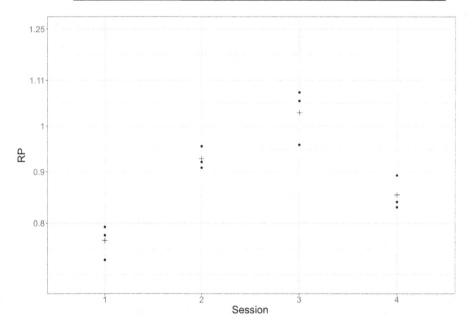

FIGURE 3.9

Plot of example data set for $n = 4$ sessions, $k = 3$ method runs per session. The geometric mean RP for each run is displayed as a +. There is greater variation betweeen the geometric means, than between points within a session.

Estimate of the between sessions variance:

$$\hat{\sigma}^2_{\text{Between sessions}} = \left(\frac{0.04323 - 0.00201}{3} \right)$$

$$= 0.01374.$$

TABLE 3.4

Analysis of variance table for the example data presented in Table 3.3. DF = degrees of freedom, SS = sum of squares, MSS = mean sum of squares and $\mathbb{E}$[MSS] = expected value of mean sum of squares.

Source	DF	SS	MSS	$\mathbb{E}$[MSS]
Between sessions	3	0.12969	0.04323	$3\sigma^2_{\text{Between session}} + \sigma^2_{\text{Within session}}$
Within session	8	0.01610	0.00201	$\sigma^2_{\text{Within session}}$
Total	11	0.14579		

Therefore, the estimate of the variance of an individual measurement across sessions of the method (on the log scale) is:

$$\hat{\sigma}^2_{\log \text{RP}} = \hat{\sigma}^2_{\text{Between sessions}} + \hat{\sigma}^2_{\text{Within session}}$$
$$= 0.01374 + 0.00201$$
$$= 0.01575.$$

The upper (one-sided) 95% confidence limit for the variance is 0.12424 and can be calculated as follows. Since we are estimating the IP, $n_F = 1$ and $k_F = 1$, this implies that both a and b are also 1. The values of H_1 and H_2 are calculated as:

$$H_1 = \frac{4-1}{\chi^2_{0.05,(4-1)}} - 1$$
$$= 7.526,$$

and:

$$H_2 = \frac{4(3-1)}{\chi^2_{0.05,4(3-1)}} - 1$$
$$= 1.928.$$

Then, the upper confidence limit is:

$$0.01575 + \sqrt{7.526^2 \left(\frac{1}{3}\right)^2 0.04323^2 + 1.928^2 \left(1 - \frac{1}{3}\right)^2 0.00201^2}$$
$$= 0.01575 + 0.10849$$
$$= 0.12424.$$

Substituting the estimates of the variance components into the formula for the IP, the estimate of the IP is:

$$\widehat{\text{IP}} = 100 \times \text{antilog}\left(\sqrt{0.01575} - 1\right) \%$$
$$= 13.4\%, \tag{3.8}$$

TABLE 3.5
Estimate of $\sigma^2_{\log \text{RV}}$ and its upper (one-sided) 95% confidence limit (UCL) for a range of values of k_F and n_F if the between and within session variances remain as 0.01374 and 0.00201, respectively.

| | | | Number of sessions, n_F | | | | |
			1	2	3	4	8	12
	1	Estimate	0.0158	0.0079	0.0053	0.0039	0.0020	0.0013
		UCL	0.1242	0.0621	0.0414	0.0311	0.0155	0.0104
Number of method runs per session, k_F	2	Estimate	0.0147	0.0074	0.0049	0.0037	0.0018	0.0012
		UCL	0.1232	0.0616	0.0411	0.0308	0.0154	0.0103
	3	Estimate	0.0144	0.0072	0.0048	0.0036	0.0018	0.0012
		UCL	0.1229	0.0614	0.0410	0.0307	0.0154	0.0102

with upper (one-sided) 95% confidence limit of 42.3%.

Using the Satterthwaite approach for comparison, the degrees of freedom are:

$$df = \frac{0.01575^2}{\frac{\left[\frac{1}{3}\times 0.04323\right]^2}{3} + \frac{\left[\left(1-\frac{1}{3}\right)\times 0.00201\right]^2}{8}}$$
$$= 3.573.$$

The upper (one-sided) 95% confidence limit for the variance in this case is:

$$\widehat{\sigma}^2_{\log \text{RV}} \times \frac{3.573}{\chi^2_{0.05,3.573}}$$
$$= 0.01575 \times \frac{3.573}{0.5466}$$
$$= 0.1030.$$

The corresponding (one-sided) 95% confidence limit for the IP is 37.8%. If a conservative approach is taken and $df = 3$ is used, the upper confidence limits are 0.1343 and 44.3% for the variance and IP, respectively.

Equation (3.4) gives the formula for the variance of the reportable value for a format with n_F repeats, each including k_F runs. Substituting in the estimates for the between and within session variances:

$$\hat{\sigma}^2_{\log \text{RV}} = \frac{0.01374}{n_F} + \frac{0.00201}{k_F \times n_F}.$$

Table 3.5 shows the estimate of $\sigma^2_{\log \text{RV}}$ and its upper (one-sided) 95% confidence limit for a range of values of k_F and n_F if the between and within session variances remain as 0.01374 and 0.00201, respectively. Table 3.6 shows the estimate expressed as a GCV along with its upper (one-sided) 95% confidence limit for the same range of values of k_F and n_F. These calculations were done without rounding the variance components.

For this example, increasing the number of sessions has a stronger effect than increasing the number of methods runs per session, for a given number of total runs. This can be

TABLE 3.6

Estimate of the GCV for the reportable value and its upper (one-sided) 95% confidence limit (UCL) for a range of values of k_F and n_F if the between and within session variances remain as 0.01374 and 0.00201, respectively.

			Number of sessions, n_F					
			1	2	3	4	8	12
	1	Estimate	13.4%	9.3%	7.5%	6.5%	4.5%	3.7%
		UCL	42.3%	28.3%	22.6%	19.3%	13.3%	10.7%
Number of method runs per session, k_F	2	Estimate	12.9%	9.0%	7.3%	6.3%	4.4%	3.6%
		UCL	42.0%	28.2%	22.5%	19.2%	13.2%	10.7%
	3	Estimate	12.8%	8.9%	7.2%	6.2%	4.3%	3.5%
		UCL	42.0%	28.1%	22.4%	19.2%	13.2%	10.6%

explained by the fact that the method runs (within a session) are more similar than those between sessions: the variance is about 7 times lower within a session; the standard deviation is about 3 times lower. There is little benefit in increasing the method runs per session beyond 3.

If the GCV is required to be at most 10%, then 2 sessions, each of one method run, are adequate. If the GCV is required to be at most 5% then 8 sessions, each of one method run, are adequate. A safer approach would be to consider the upper confidence limit for GCV. In that case, 14 and 53 sessions, respectively, would be needed (data not shown): with data available for relatively few runs, the confidence intervals are very wide.

3.4.3 Determining the format for the bioassay release procedure

The bioassay release procedure format should be designed based on the ATP for the bioassay. The ATP may driven by the proportion of manufactured lots that are expected to fall out of specification, and this is driven by business need. We assume that the limit on the proportion of out-of-specification results is 1%.

From Chapter 2, Equation (2.16):

$$Pr(\text{OOS}) = \Phi\left\{\frac{[\log(\text{LSL}) - \mu_{\text{Manuf}}]}{\sqrt{\sigma^2_{\text{Manuf}}}}\right\} + 1 - \Phi\left\{\frac{[\log(\text{USL}) - \mu_{\text{Manuf}}]}{\sqrt{\sigma^2_{\text{Manuf}}}}\right\},$$

where:

- μ_{Manuf} is the mean bioassay-reported relative potency (log scale)

    - $\mu_{\text{Manuf}} = \log\left(\frac{\text{RB}}{100} + 1\right)$ when the process is centred around RP = 1

- $\sigma^2_{\text{Manuf}} = \frac{\left[\log\left(\frac{\text{GCV}}{100} + 1\right)\right]^2}{(1-P)}$

    - GCV is the format variability

TABLE 3.7

Estimated proportion of out-of-specification results for a range of values of k_F and n_F assuming that RB $= -10.8\%$ (rounded) and the variance components (unrounded), $P = 0.2$, and specification limits of $(0.70, 1.43)$.

		Number of sessions, n_F			
		1	2	3	4
Number of method	1	4.24%	0.73%	0.14%	0.03%
runs per session,	2	3.74%	0.58%	0.10%	0.02%
k_F	3	3.57%	0.53%	0.09%	0.02%

– P is the proportion of the total variance of the reportable value that is accounted for by the manufacturing process.

For example, suppose the bioassay relative bias is estimated to be -10.8% (as calculated in Chapter 2) and the IP is estimated to be 13.4% (as calculated in Equation 3.8). Using these rounded values, suppose that the proportion of the total variance that is due to the bioassay is 0.8 and the remaining 0.2 is due to manufacturing variability. Finally, suppose that the manufacturing process is centred at RP $= 1$ and that the specification limits are $(0.70, 1.43)$. If the release format consists of a single run of the method, then, assuming the estimates of RB and IP are the true values, the proportion of lots from the manufacturing process that is expected to fall out of specification is given by:

$$
Pr(\text{OOS}) = \Phi \left\{ \frac{\left[\log(0.70) - \log\left(\frac{-10.8}{100} + 1\right)\right]}{\sqrt{\frac{\left[\log\left(\frac{13.4}{100} + 1\right)\right]^2}{(1-P)}}} \right] + 1 - \Phi \left\{ \frac{\left[\log(1.43) - \log\left(\frac{-10.8}{100} + 1\right)\right]}{\sqrt{\frac{\left[\log\left(\frac{13.4}{100} + 1\right)\right]^2}{(1-P)}}} \right\}
$$

$$
= \Phi \left\{ \frac{\left[\log(0.70) - \log(0.892)\right]}{\sqrt{\frac{\left[\log(1.134)\right]^2}{0.8}}} \right\} + 1 - \Phi \left\{ \frac{\left[\log(1.43) - \log(0.892)\right]}{\sqrt{\frac{\left[\log(1.149)\right]^2}{0.8}}} \right\}
$$

$$
= 4.27\%.
$$

Therefore, a single run of the method is expected to result in almost 5% of manufactured lots falling out of specification, so the assay format will require more replication. Table 3.7 provides the values of proportion of out-of-specification results corresponding to the same set of values of n_F and k_F as in Tables 3.5 and 3.6.

With two sessions of one method run each, i.e., $n_F = 2$, $k_F = 1$, the proportion of out-of-specification results is 0.73%. However, given that the values for RB and GCV that were used were estimated based on only 12 data points (arising from 4 sessions), it would be risky to set the release procedure format on these figures.

Table 3.8 shows the proportion of out-of-specification results with more conservative (arbitrary) values: RB is set to -16.7% and IP is set to 20%. Note that for formats with $k_F > 1$, the calculation is more complex as it depends on the individual variance components. Assuming $k_F = 1$ gives a conservative estimate of proportion of out-of-specification results; adding more runs per session will improve the precision.

TABLE 3.8

Estimated proportion of out-of-specification results for RB $= -16.7\%$, GCV $= 20\%$, $P = 0.2$ and specification limits of $(0.70, 1.43)$ for different numbers of sessions, each consisting of 1 run.

	Number of sessions, n_F					
	1	2	3	4	6	8
1 run per session	20.1%	11.4%	7.0%	4.4%	1.8%	0.8%

A safer option would be to choose $n_F = 8$, $k_F = 1$, providing an estimated proportion of out-of-specification results of less than 1% when the relative bias is -16.7% (equivalent to 20% on the positive side – see Section 2.3.1.1) and the intermediate precision is 20%. Assuming that 1% is an acceptable out-of-specification proportion, the ATP could reasonably include limits of $(-17\%, 20\%)$ for the RB and 20% for the IP, assuming that the release procedure is to include 8 sessions of one run each. However, including this number of replicates might introduce additional levels of variability which were not captured in the estimates used.

3.5 Chapter summary

- This chapter covers the design and optimisation of a bioassay with a view to achieving the required performance in a given application.

- The layout of samples and doses in wells on plates (or animals in cages), the number of independent replicates to be included at each dose, and the number of plates to be included in the bioassay method, are considered.

- The design of the plate layout is important to both the accuracy and precision of the method, while the replication strategy for the assay method further contributes to precision.

- Between-run and within-run variance are described. Background responses on the plate are evaluated and designs which avoid the potential bias introduced by patterns in the background are discussed.

- The use of quality control samples is discussed. The design of the release procedure, which is driven by the analytical target profile for the bioassay, is also explored.

- The discussion applies equally to relative potency assays and interpolation-type assays, where a concentration is estimated.

Part II

Estimation of potency: the statistical analysis of bioassay runs

4

Statistical models for characterising the dose-response relationship

In order to calculate the relative potency, or concentration, for a test sample we need to find a mathematical form for a dose-response relationship that fits the assay data. The relationship will be defined by a functional form which involves a number of parameters – that is, values that fit the curve to the data. For example, the functional form which is a straight line has two parameters that define it: the slope and the intercept. In most cases the response will be a function of the logarithm of the dose.

In the present chapter we consider the relationship between the dose and the true mean response: that is, we ignore the variability in the data. In the following chapter we will consider the variability of the data around that mean response for a given dose. The focus is on relative potency assays. For interpolation-type assays, the same considerations apply for the dose-response relationship for the standard curve.

4.1 Response types

We discuss two main categories of response data: continuous and binary. Continuous responses can take any value (within a given range) for an individual measurement. Binary responses can only take one of two possible values, usually denoted by 0 and 1. Binary data are sometimes referred to as 'quantal' in the bioassay context. These two categories of response data need different statistical approaches – firstly because the dose-response relationships have different characteristics and secondly because the models for their variability are different (see Chapter 6).

When the recorded response is binary, it usually arises from an underlying continuous measure. For example, for a binary response of 'dead or alive at the end of the study', the underlying variable would be 'time to death'. Subjects are observed up to the end of the study. If they die during the study, the value of 'time to death' is observable and is less than the length of the study, and the recorded binary response is 'dead' or 1. If they survive to the end of the study the value of 'time to death' is not observable but is known to be at least the length of the study. In this case the recorded binary response is 'alive' or 0. Time to event data such as these can be analysed using techniques commonly used in clinical

DOI: 10.1201/9781003449195-4

trials, engineering and economics which make much more efficient use of the data available than the use of the dichotomised binary variable. The data analysis handles the fact that the continuous response value may not be known exactly.

4.2   Shape of dose-response curve

A general assumption is that the full underlying dose-response relationship is an S-shaped (or sigmoid) curve, where the response is either ever-increasing or ever-decreasing as the dose increases from zero. At each end of the curve the response flattens and approaches a limit. The response limits, corresponding to zero and infinite doses, are called asymptotes. Such a dose-response relationship is shown in Figure 4.1(a).

In some assay systems the response can reach a maximum and then start to reduce (a 'hook' effect) at the upper end of the response range. In these cases the standard mathematical forms are not applicable and the usual solution is to restrict the doses used to the range where the selected relationship applies.

The central part of the S-shaped dose-response is approximately linear, and some assays are designed to restrict responses to only the linear part of the curve. This should be avoided if practically possible. Figure 4.1(b) shows the dose-response relationship for just the central doses of the full S-shaped dose-response curve shown in Figure 4.1(a). Figure 4.2 then shows the dose-response curve at these central doses for a reference and test sample pair at different relative potencies. In all cases, the true underlying dose-response curves are parallel. However, when looking at the relationship over only the central doses, they exhibit non-parallelism. As the relative potency moves away from 1, the curves separate (horizontally), and the fitted straight lines become less parallel. Estimation of the RP in these cases will be poor.

4.3   Continuous data

We begin our discussion with the case of a continuous response measurement. A continuous response can take on any value (within a range). Examples include luminescence and optical density for cell-based assays, where the observed response can be any value within the limits of the measurement device. We will examine the mathematical forms for both straight-line and S-shaped relationships.

4.3.1   Straight-line relationship

The simplest dose-response relationship is a straight line. The straight-line relationship is often a convenient simplification of a more complicated dose-response relationship. Therefore,

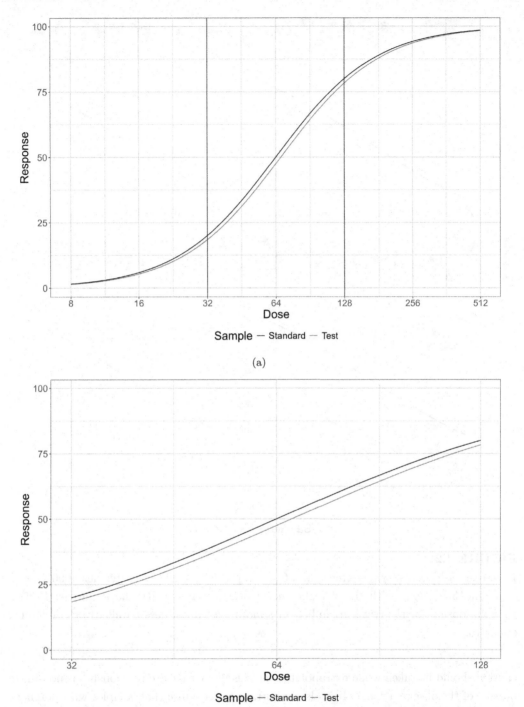

(a)

(b)

FIGURE 4.1

S-shaped dose-response relationship in (a); an approximately linear dose-response curve over the central doses in (b).

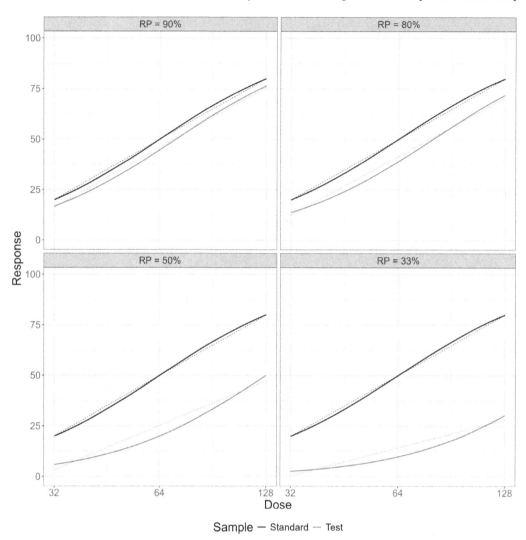

FIGURE 4.2

Dose-response curves for a reference and test sample pair at relative potencies of 90%, 80%, 50%, and 33% along with the straight-line of best fit. As the RP moves further from 1, the relationship for the test sample becomes more curved and less parallel to the reference standard.

caution should be taken when extrapolating as it is likely that the relationship is non-linear outside of the observed range. It is also important to recognise that samples with potencies far from 1 may fall outside the linear range (see Figure 4.2).

4.3.1.1   Log-transformed dose

Most commonly, the dose-response relationship is linear with respect to the log-transformed dose. That is, we expect the response to change in an additive fashion when the dose is changed multiplicatively. If we also expect the response to change multiplicatively, then the

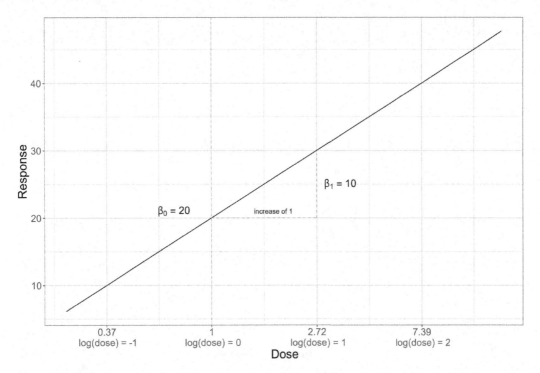

FIGURE 4.3

Straight-line relationship with respect to the log-transformed dose. For every one unit increase in the log dose, the response changes by $\beta_1 = 10$ units.

response can also be log-transformed to achieve a straight line relationship.

Let $\mu_{Y|\text{dose}}$ be the mean response at a given dose. Then, the relationship between the mean response and the dose can be stated as:

$$\mu_{Y|\text{dose}} = \beta_0 + \beta_1 \times \log(\text{dose}),$$

where:

β_0 is the intercept, and corresponds to the mean response value when $\log(\text{dose}) = 0$ or equivalently at a dose $= 1$,

β_1 is the slope, and corresponds to the change in μ_Y for an increase of 1 in $\log(\text{dose})$ or when the dose is multiplied by a factor of antilog(1).

Figure 4.3 illustrates the straight-line relationship between response and the log-transformed dose for intercept, $\beta_0 = 20$, and the slope, $\beta_1 = 10$.

Parallel lines

When estimating a relative potency, we have a reference standard and a test sample whose dose-response relationships are parallel. For a straight-line relationship, parallelism occurs when the slope parameter, β_1, is the same for both samples. Figure 4.4 shows a parallel

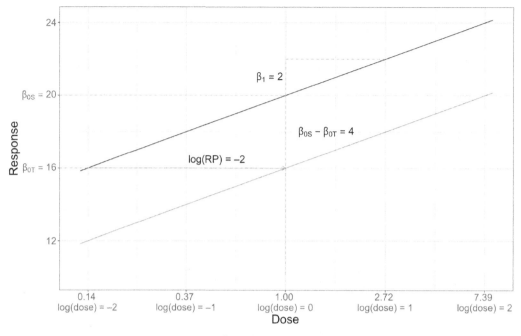

FIGURE 4.4

Parallel straight-line relationship with respect to the log-transformed dose for a reference standard and test sample pair. When the lines are parallel the horizontal distance between the two lines is constant at $-\log(\mathrm{RP}) = 2$.

straight-line dose-response relationship for a reference standard and test sample pair. If the relative potency is 1, then the intercepts will be identical; otherwise they will differ. Let the subscripts S and T represent the standard and test sample, respectively. Then, the parallel dose-relationships for the two samples are:

Standard:

$$\mu_{Y|\text{dose},S} = \beta_{0S} + \beta_1 \times \log(\text{dose}),$$

Test:

$$\mu_{Y|\text{dose},T} = \beta_{0T} + \beta_1 \times \log(\text{dose}),$$

where β_{0S} and β_{0T} are the sample specific intercepts and β_1 is the shared (common) slope.

When the relationship is linear with respect to the log-transformed dose, then the horizontal distance between the lines associated with the reference standard and test sample corresponds to the log-transformed relative potency. The slope can be written as rise, calculated as the difference in the intercepts (standard - test), over run (negative log RP):

$$\beta_1 = \frac{(\beta_{0S} - \beta_{0T})}{-\log(\mathrm{RP})}.$$

Therefore,

$$\log \mathrm{RP} = \frac{(\beta_{0T} - \beta_{0S})}{\beta_1}$$

and

$$\mathrm{RP} = \mathrm{antilog}\left[\frac{(\beta_{0T} - \beta_{0S})}{\beta_1}\right].$$

For the example shown in Figure 4.4, the RP is:

$$\mathrm{RP} = \mathrm{antilog}\left[\frac{(16 - 20)}{2}\right]$$

$$= 0.14.$$

4.3.1.2   Untransformed dose

For some assay systems, the dose-response relationship will be linear with respect to the (untransformed) dose. That is, an additive change to the dose, results in an additive change to the mean response, μ_Y. This relationship can be stated as:

$$\mu_{Y|\mathrm{dose}} = \beta_0 + \beta_1 \times \mathrm{dose},$$

where:

β_0 is the intercept and corresponds the predicted response value when dose $= 0$,

β_1 is the slope and corresponds to the change in μ_Y for an increase of 1 in dose.

Recall the definition of RP is the ratio of doses that result in the same response:

$$\mathrm{RP} = \frac{\mathrm{dose}_S}{\mathrm{dose}_T}.$$

For a given response, R, the dose required can be calculated from the line. For the reference standard:

$$R = \beta_{0S} + \beta_{1S} \times \mathrm{dose}_S$$

Therefore,

$$\mathrm{dose}_S = \frac{R - \beta_{0S}}{\beta_{1S}}.$$

Similarly, for the test sample:

$$dose_T = \frac{R - \beta_{0T}}{\beta_{1T}}.$$

Therefore,

$$\mathrm{RP} = \frac{(R - \beta_{0S}) \times \beta_{1T}}{(R - \beta_{0T}) \times \beta_{1S}}.$$

Unless the intercepts are equal, $\beta_{0S} = \beta_{0T}$, this ratio depends on the response level. Therefore, for the RP to be uniquely defined, the intercepts must be the same. This condition replaces the condition of parallelism which applies when the dose-response relationship is based on the log scale for the dose.

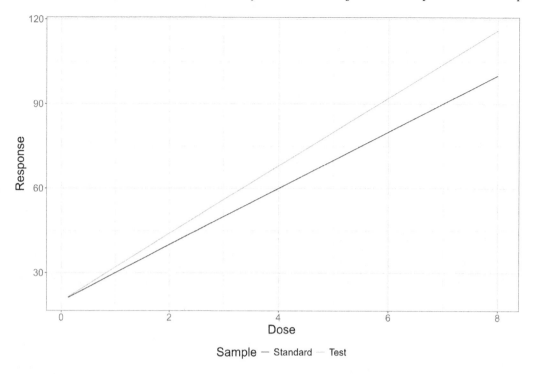

FIGURE 4.5
Straight-line relationship with respect to the original scale dose for a reference standard and test sample pair. The lines are not parallel but have the same intercept.

When $\beta_{0S} = \beta_{0T} = \beta_0$, the dose-response relationships for the two samples are:

Standard:
$$\mu_{Y|\text{dose},S} = \beta_0 + \beta_{1S} \times \text{dose},$$

Test:
$$\mu_{Y|\text{dose},T} = \beta_0 + \beta_{1T} \times \text{dose},$$

where β_0 is the shared (common) intercept and β_{1S} and β_{1T} are the sample-specific slopes. The RP can then be calculated as the ratio of the slopes:

$$\text{RP} = \frac{\beta_{1T}}{\beta_{1S}}.$$

For this reason, when there is a straight-line relationship between the untransformed dose and the response the term 'slope-ratio model' is often used. Figure 4.5 shows this relationship for a reference and test pair.

As a special case, it is interesting to note that, when the common intercept is zero, the dose-response relationships are:
Standard:
$$\mu_{Y|\text{dose},S} = \beta_{1S} \times \text{dose},$$

Test:

$$\mu_{Y|\text{dose},T} = \beta_{1T} \times \text{dose}.$$

The ratio of the test sample response to the reference standard response is:

$$\frac{\mu_{Y|\text{dose},T}}{\mu_{Y|\text{dose},S}} = \frac{\beta_{1T}}{\beta_{1S}},$$

and corresponds to the RP. This is the only scenario where the ratio of the responses can be used to calculate the RP.

This relationship is mentioned for completeness but not developed further in this book.

4.3.2 S-shaped relationships

Typically, the observed dose-response relationship for continuous responses is non-linear with limiting lower and upper responses. Describing such data mathematically is inherently more complicated, and requires more parameters than the straight-line relationship previously discussed. In this section, we will discuss a family of logistic curves with 3 to 5 parameters. Increasing the number of parameters allows for the curves to fit increasingly complex shapes.

4.3.2.1 4-parameter logistic relationship

The most commonly used non-linear relationship is the symmetric, S-shaped 4-parameter logistic curve (4PL) shown in Figure 4.6. There are many different formulations for this curve that are mathematically equivalent. However, for this book we will use the following:

$$\mu_{Y|\text{dose}} = D + \frac{A - D}{1 + e^{B[\log(\text{dose}) - C]}},$$

For positive values of B:

A is the left asymptote (dose $= 0$),

B is the slope parameter (sometimes called the effect rate),

C is the horizontal location of the curve,

D is the right asymptote (dose $= \infty$).

The asymptotes, A and D, correspond to the limiting responses of the assay. Note that if the value of B is negative, then the asymptotes are reversed, and A corresponds to the right asymptote, D the left asymptote.

We refer to B as the 'slope parameter' rather than the 'slope'. There is no single 'slope'. With reference to Figure 4.6, starting at the left-hand side of the plot, the actual slope of the 4PL ranges from zero and increases until it reaches the inflection point at C when it then begins to decrease again before returning to zero. The value of the slope at the midpoint of the curve (where $\mu_Y = \frac{(A+D)}{2}$ at $\log(\text{dose}) = C$) can be shown (via calculus) to be:

$$\text{Slope at midpoint} = \frac{B(D - A)}{4}.$$

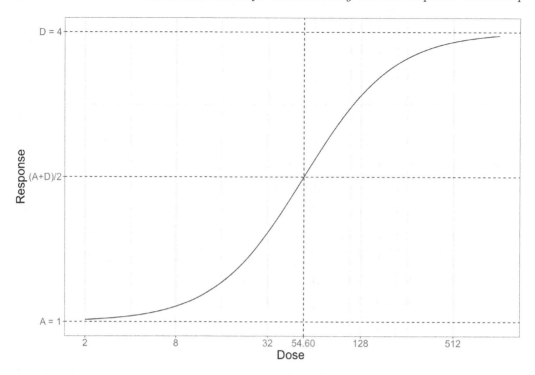

FIGURE 4.6

Symmetric, S-shaped 4-parameter logistic (4PL) curve. The parameters are: $A = 1$, $B = 1.5$, $C = \log(54.60) = 4$, and $D = 4$.

Therefore, B can be interpreted as a 'normalised' value of the slope at the midpoint:

$$B = \text{Slope at midpoint} \times \frac{4}{(D - A)}.$$

For a given pair of values A and D, higher values of B correspond to steeper slopes at the midpoint.

Finally, the dose at the midpoint of the curve is the EC_{50}. That is, the dose at which a half-maximal response is observed. Therefore, the parameter C can be expressed as $\log(EC_{50})$ and antilog$(C) = EC_{50}$. Note that when the curve is decreasing, the midpoint is sometimes called the inhibitory concentration, IC_{50}.

Parallel curves

As for the straight-line relationship, estimation of the relative potency from the 4PL relationship requires parallelism. Parallelism occurs when the curves associated with the reference standard and test sample share the same values for the asymptotes and slope parameter and differ only in their midpoint. That is, the parameters A, D and B for the standard and test sample are required to be the same for parallelism to be met. If any one (or more) of these pairs of parameters are not equal, then the curves are not parallel and the relative potency is not unique. Figure 4.7 illustrates the 4PL relationship for a pair of parallel standard and test curves.

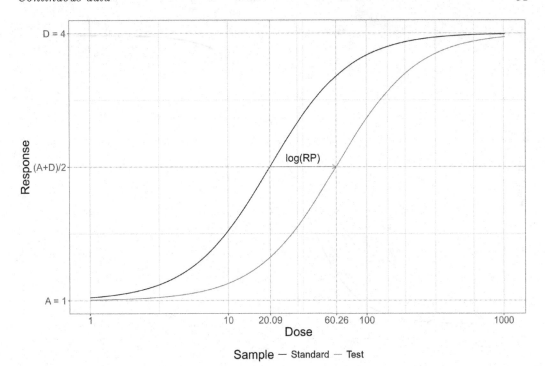

Sample — Standard — Test

FIGURE 4.7

Parallel symmetric, S-shaped 4-parameter logistic (4PL) curves. The horizontal distance between the curves is constant. The parameters are: $A = 1$, $B = 1.5$, $C_S = \log(20.09) = 3$, $C_T = \log(60.26) = 4.10$, $D = 4$ and RP $= 0.33$.

The parallel dose-relationships for the two samples are:

Standard:

$$\mu_{Y|\text{dose},S} = D + \frac{A - D}{1 + e^{B(\log(\text{dose}) - C_S)}},$$

Test:

$$\mu_{Y|\text{dose},T} = D + \frac{A - D}{1 + e^{B(\log(\text{dose}) - C_T)}},$$

where parameters A, B, and D are shared between the two samples and C_S and C_T are the sample-specific $\log(\text{EC}_{50})$. The horizontal distance between the curves can be measured as the difference between the C parameters and so the $\log$ RP is:

$$\log \text{RP} = C_S - C_T.$$

Therefore:

$$\text{RP} = \frac{\text{antilog}\,(C_S)}{\text{antilog}\,(C_T)}$$

$$= \frac{\text{Standard EC}_{50}}{\text{Test EC}_{50}}.$$

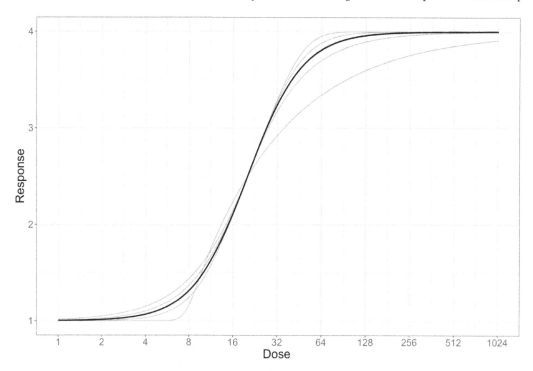

FIGURE 4.8

Asymmetric, S-shaped 5-parameter logistic (5PL) curves over a range of E parameters. When $E = 1$ the curve is symmetric and is equivalent to the 4PL curve (dark black curve).

It is important to note that these EC_{50} values come from parallel curves and so their difference measures the (constant) horizontal distance between the curves. When the curves are not parallel, the difference between the EC_{50}s will not, in general, be at the same response level, and so their difference will not be a measure of the horizontal distance between the curves. Consequently, it is not appropriate to use EC_{50} values from non-parallel relationships in the calculation of RP.

4.3.2.2 5-parameter logistic relationship

The 4PL is a symmetric, S-shaped relationship which, whilst very common, does not always reflect the dose-response relationship seen in the data. The dose-response relationship for some assays will be asymmetric. That is, the curve will be steeper (or shallower) at the top when compared to the bottom. An additional parameter can be introduced to capture a degree of asymmetry in the relationship. The 5-parameter logistic (5PL) is an example:

$$\mu_{Y|\text{dose}} = D + \frac{A - D}{\left\{1 + e^{B[\log(\text{dose}) - C]}\right\}^E}.$$

The (always positive) parameter E controls the extent of the asymmetry. As a special case, when $E = 1$ the relationship is symmetrical and becomes the 4PL. Figure 4.8 illustrates the 5PL relationship for a range of values of E.

As for the 4PL, $\mu_{Y|\text{dose}}$ is the mean response for a given dose, and A and D are the asymptotes. The parameter E expresses the amount of asymmetry $(E > 0)$. The interpretations of B and C are more complex. When the log dose is C:

$$\mu_Y = D + \frac{(A - D)}{2^E}.$$

The value of the slope when the log dose is C is:

$$\text{Slope} = \frac{B(D - A)}{2^{(1+E)}}.$$

Therefore, B can be interpreted as a 'normalised' value of the slope when log dose $= C$ in a similar way to the 4PL slope parameter. The $\log \text{EC}_{50}$ is given by:

$$\log(EC_{50}) = C + \frac{1}{B}\log(2^{\frac{1}{E}} - 1), \tag{4.1}$$

with the EC_{50} found by taking the antilog of Equation (4.1). When $E = 1$, $C = \log(\text{EC}_{50})$, as for the 4PL.

Parallel curves

Parallelism for the 5PL model occurs when the asymmetry parameters, E, for the standard and test curves are equal, in addition to the asymptotes and slope parameter. That is, the parallel dose-relationships for the two samples are:

Standard:

$$\mu_{Y|\text{dose},S} = D + \frac{A - D}{\left[1 + e^{B(\log(\text{dose}) - C_S)}\right]^E},$$

Test:

$$\mu_{Y|\text{dose},T} = D + \frac{A - D}{\left[1 + e^{B(\log(\text{dose}) - C_T)}\right]^E}.$$

Figure 4.9 illustrates the 5PL relationship for a standard and a test sample. The two curves share values of A, B, D, and E. Therefore, although they do not appear to be, the two curves are parallel.

As for the 4PL model, the horizontal distance between the curves can be measured by the difference in the C parameters for the standard and test samples. That is, the $\log \text{RP}$ is:

$$\log \text{RP} = C_S - C_T.$$

Therefore:

$$\text{RP} = \frac{\text{antilog}(C_S)}{\text{antilog}(C_T)}.$$

4.3.2.3   3-parameter logistic relationship

If the limiting value of the response of the 4PL at either end of the dose range is known, it could be argued that the asymptote value should be fixed rather than estimated from

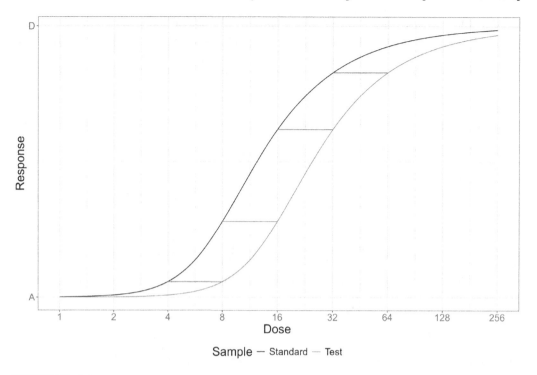

FIGURE 4.9

Parallel asymmetric, S-shaped 5-parameter logistic (5PL) curves. The horizontal distance is constant.

the data. For example, where the response is a percentage of cells killed it might be argued that, at a (theoretically) infinite dose, 100% of cells would die. In this case one might fit a logistic relationship to the data where the EC_{50}, slope parameter and one asymptote are the three unknown parameter values. This is known as a 3-parameter logistic relationship.

The problem with fixing one of the asymptotes is that the assumption underlying it may be incorrect and the resulting model may not fit the data well. This can cause a bias in the estimate of potency. In addition, fixing the asymptote can cause an under-estimate of the variance and a spurious narrowing of the confidence interval. Fitting a 4PL relationship to the data is recommended in this situation, and we do not discuss the 3PL relationship in any further detail.

4.4 Binary data

When the response for an individual subject is binary: an event occurs or it does not occur. For example, the subject is dead (response = 1) or alive (response = 0) at the end of the assay. In this case, each subject has a probability of death which is related to the dose given.

As for continuous data, the dose-response relationship is assumed to be S-shaped. In this case however, the data have natural limits and the asymptotes will be 0 and 1 and need not be estimated from the data. There are two mathematical relationships that are commonly used in this situation: the logit function and the probit function. These are very similar in form and, in practice, it makes little difference which is chosen. We do not discuss the probit function further.

4.4.1 Logit relationship

When the response for each subject is binary the relationship between dose and response is expressed in terms of the probability, p, that the response is 1 when the subject receives a given dose:

$$p = \frac{1}{1 + e^{-(\beta_0 + \beta_1 \times \log(\text{dose}))}},$$

where β_0 is the intercept parameter and β_1 is the slope parameter. For positive values of β_1, at zero dose, $p = 0$ and at infinite dose, $p = 1$. For negative values of β_1, these are reversed.

This functional relationship is the same as the 4PL with asymptotes at 0 and 1. It is important to note that the assumptions that underlie the analysis of binary data are different from the continuous data case this will be discussed further in Chapter 5.

This relationship can also be written as:

$$\log\left(\frac{p}{1-p}\right) = \beta_0 + \beta_1 \times \log(\text{dose}).$$

The left-hand side of the equation is commonly referred to as the logit of p and has a straight-line relationship with respect to the log dose. The ED_{50} is the dose resulting in $p = 0.5$:

$$\log\left(\text{ED}_{50}\right) = \frac{-\beta_0}{\beta_1}.$$

Parallel lines

For the logit relationship, parallelism occurs when the slope β_1 is equivalent for a standard and test sample pair. Figure 4.10 illustrates a pair of parallel logit dose-response curves. If the β_0 parameters for the standard and test samples are denoted by β_{0S} and β_{0T}, respectively, the horizontal distance between the curves, $\log \text{RP}$, is related to β_{0S}, β_{0T} and β_1 as follows:

$$\log \text{RP} = \frac{-\beta_{0S}}{\beta_1} - \frac{-\beta_{0T}}{\beta_1}$$

$$= \frac{(\beta_{0T} - \beta_{0S})}{\beta_1},$$

with

$$\text{RP} = \frac{\text{Standard ED}_{50}}{\text{Test ED}_{50}}.$$

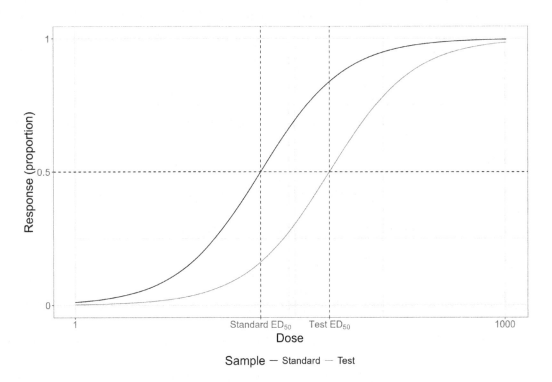

FIGURE 4.10

Parallel logit dose-response relationship. The response here is a proportion and ranges from 0 to 1. The ED_{50}s correspond to a response proportion of 0.5. The horizontal distance between the curves is constant.

4.5 Chapter summary

- The mathematical models typically used to describe dose-response relationships in bioassays are introduced.

- For continuous responses, the straight-line, the 4-parameter logistic (4PL) and the 5-parameter logistic (5PL) relationships are covered.

- The use of the log(dose) is explained. The mathematical forms (parameterisations) and the risks and benefits associated with the model choice are outlined.

- The use of the slope ratio model when the dose-response relationship is linear with respect to the dose is explored.

- For binary data, the logit model is explained.

- For all models, the interpretation of the model parameters is provided analytically and via graphics. The relationship between the model parameters and the relative potency is provided.

5

Choosing the dose-response relationship for bioassay data

This chapter deals with the choice of dose-response relationship that will be used in routine analysis for the assay method, including checking that the required assumptions hold for the data and the possible transformation of the response.

The choice of dose-response relationship to be used for a bioassay involves several considerations. Firstly, the fitted relationship needs to match the data points. Data points that lie far from the fitted curve (or line) could be technical outliers, arising from human error or equipment malfunction, or they could be valid data points. Technical outliers should be deleted from the data set. Valid data points should be retained; they may influence the choice of dose-response relationship. Deciding whether a data point is 'far from the fitted curve' is a statistical issue which we address in Chapter 6; deciding whether an outlier is technical or not is the scientist's responsibility.

Secondly, when making inferences about the test sample's relative potency, or concentration, from the bioassay data, we need an assessment of the confidence we have in the result. This assessment usually takes the form of a confidence interval and requires that a number of assumptions about the data are met. These assumptions need to be checked to ensure that the calculations are valid. The assumptions also apply to the suitability tests described in Chapter 7.

5.1 Continuous data: statistical assumptions

The assumptions made in bioassay analysis for continuous responses are that they have the following attributes:

- Independence among the observations;

- Normal distribution:

    - mean value dependent on the dose;

    - variance constant across the doses.

DOI: 10.1201/9781003449195-5

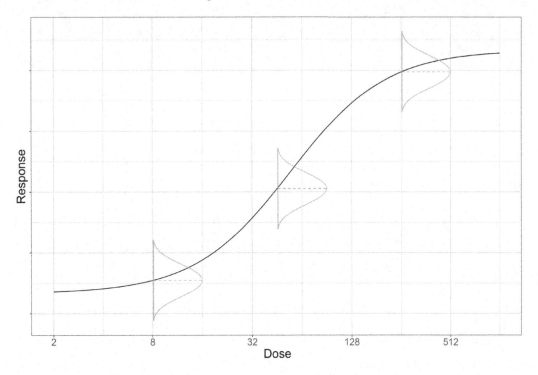

FIGURE 5.1

For each dose group, the mean response lies on the dose-response curve. The variance (the width of the bell curve) is constant (homogeneous) across all the doses.

That is, the response of a subject receiving a given dose (of a given preparation) is assumed to vary randomly, the values being distributed normally around a mean response value, with a variance which is constant across all response levels. Figure 5.1 illustrates this variation in the data. For each dose group, the mean response lies on the dose-response relationship. The variance (the width of the bell curve) is constant (homogeneous) across all the doses.

5.1.1 Constant variance

A common pattern seen in bioassay data is an increase in the variability as the response increases. This is a clear violation of the assumption of homogeneity of variance and can cause estimates of the RP and its confidence interval to be biased. For example, the data in Figure 5.2 show a pattern of increasing spread with increasing response. The data for this figure are found in Table B.1. A transformation of the responses can often achieve the required homogeneity.

Whether or not there is a problem with variance homogeneity is not always as clear as this, and it is necessary to examine data from multiple assay runs before coming to a firm conclusion about the constancy of the variance. When there are replicate responses at each dose, the following plots can help to support the assessment when based on several assay runs. In each case the mean response, calculated per dose group per run, is shown on the x axis.

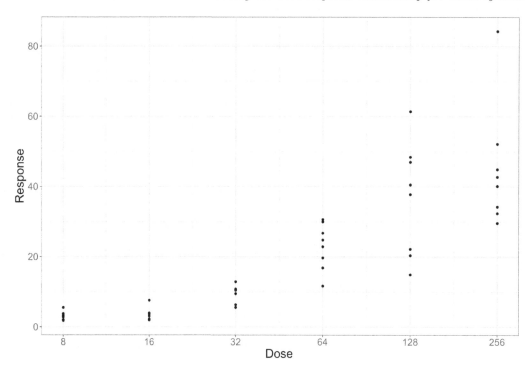

FIGURE 5.2

Assay response values for eight replicates at six dose levels. The variability (spread) of the response values increases as the assay response increases. The data for this figure are found in Table B.1.

- Residual values (data value minus mean for its dose group) versus mean response.

  – If the residuals fan out at higher response levels then a log or square root transformation of the response should be considered.

- Standard deviation (SD) of the responses versus mean response.

  – If the slope of a linear regression is positive then a log or square root transformation may be appropriate.

- Coefficient of variation (CV = SD/mean) of the responses at each dose versus mean response.

  – If the CV is stable across values of the mean then a log transformation of the response will result in approximately constant variance.

Figure 5.3 illustrate these plots for the data set in Table B.1. Figure 5.3(a) shows that the residuals are fanning out at higher response levels, while Figure 5.3(b) shows that the standard deviation increases with mean response. Finally, Figure 5.3(c) shows that the coefficient of variation is stable across the values of the mean. All of these together suggest that a log-transformed response may be appropriate for these data.

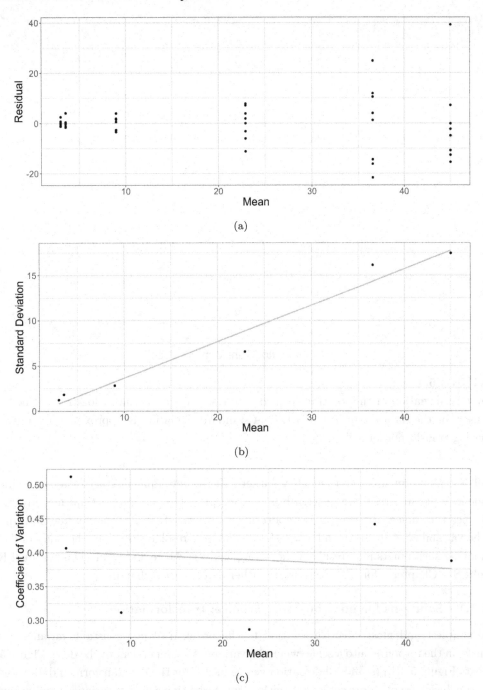

FIGURE 5.3

Mean value for each dose group plotted against: the individual residual values in (a); the standard deviation for each dose group in (b); and the coefficient of variation for each dose group in (c). The spread of the points and the standard deviation increase with respect to the mean response, the coefficient of variation is approximately constant. The data for this figure are found in Table B.1.

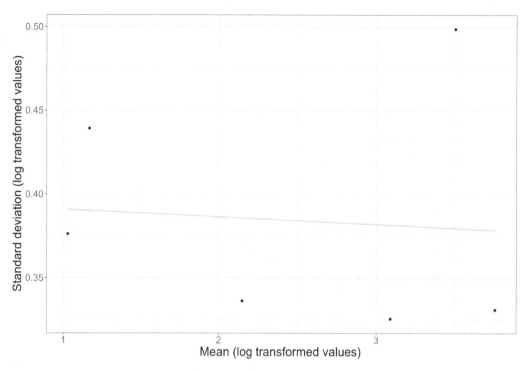

FIGURE 5.4

Standard deviation of the log-transformed responses versus the mean log-transformed responses for the data set in Table B.1. The standard deviation is approximately constant after log transformation.

These plots can also help to identify outliers. At the development stage, where the aim is to identify the dose-response relationship, it is acceptable to omit points that may adversely influence the choice. However, it is important to be aware that such points may form part of the variability of the data and a transformation may resolve them into the body of data. Removing such points may result in the wrong analysis choice and would bias the final RP result. See Chapter 6 for more details on outlier detection and handling.

5.1.1.1 Solutions for non-constant variance: transformation

The simplest and most effective solution to the problem of non-constant variance is to transform the response onto a scale, where its variance is independent of the dose. Figure 5.4 repeats Figure 5.3(b) (standard deviation versus mean) for the log-transformed data values. The conclusion for this data set is that the log transformation achieves variance homogeneity.

5.1.1.2 Solutions for non-constant variance: weighting

An alternative approach to the problem of heterogeneity of variance is to apply weights to the response data to compensate for the effect of the variability on the estimation of the parameters. The aim is to assign more weight to groups, where the variability is low and less weight when it is high. One simple approach is to apply weights to the responses that are inversely proportional to the response variance. However, this is a data dependent choice and these weights are highly variable from run to run. We do not consider weighting further in this book.

5.1.2 Normality

Let $i = 1, \ldots, d$ index the dose groups with n_i observations in dose group i. Then let $j = 1, \ldots, n_i$ index the observations within dose group i. The jth response in dose group i, Y_{ij}, will have a normal distribution with mean μ_i and variance σ^2, that is:

$$Y_{ij} \sim N(\mu_i, \sigma^2).$$

The mean, μ_i, is a function of the dose (see Chapter 4 for the dose-response relationships for μ_i); the variance, σ^2, is constant across all dose groups .

There are usually few data values per dose group, making it hard to assess their distribution. The within-group residuals, i.e., the differences between the data value and the dose group mean, can be pooled across all of the dose groups to assess the normality of the data. If a transformation is to be used to ensure homogeneity of variance, this should be applied before assessing normality.

Graphical methods can be used to assess the normality of the residuals. A histogram gives a visual impression of symmetry (which is required for normality) but does not allow easy assessment of the shape of the 'bell' curve. A normal quantile-quantile (QQ) plot displays the observed data quantiles against the theoretical quantiles assuming a normal distribution. If the data come from a normal distribution, then the points will lie roughly along on a straight line (see Appendix A for more details). A normal QQ plot allows for an easier visual assessment of normality than a histogram. Formal statistical tests can also be conducted to test if the data show evidence of non-normality. An example of such a test is the Shapiro-Wilk test [56].

Figure 5.5 plots the data in Table B.2 and shows the observed data (with the group means) in (a), a histogram of the residuals in (b) and a QQ plot of the residuals in (c). Visually we can see that both the histogram and QQ plot show skewness suggesting a lack of normality. The Shapiro-Wilk test p-value is 0.010, also indicating that the data are not consistent with the assumption of normality.

5.1.2.1 Solutions for non-normality: transformation

Where data are skewed, log-transformed data are often normally distributed. If the data arise from counts, a square root transformation is often appropriate. If the data are fundamentally proportions or percentages, the arc sine transformation may be appropriate. Normality can be re-assessed using the transformed data to confirm the transformation.

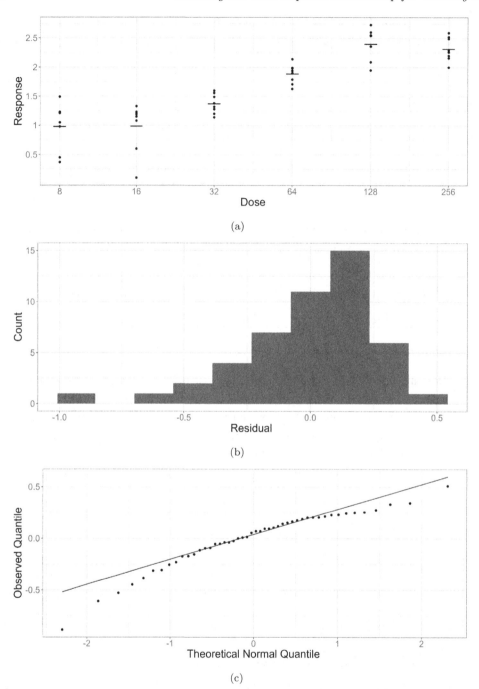

(a)

(b)

(c)

FIGURE 5.5

Response data for a set of six doses along with group mean represented by a horizontal line in (a); histogram of the residuals in (b); and QQ-plot of the residuals in (c). There is clear evidence that the data are not normally distributed. The data for this figure are found in Table B.2.

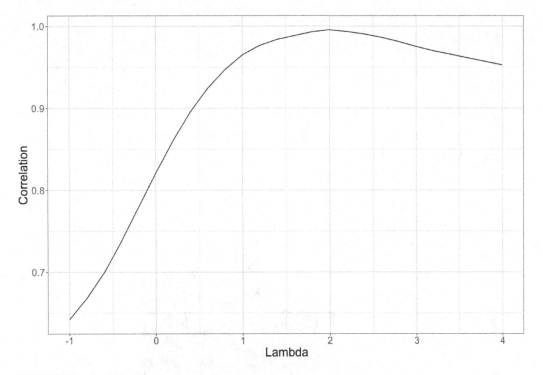

FIGURE 5.6

Correlation coefficient of the QQ plot of the transformed data versus value of λ for the data set in Table B.2. The value of λ that maximises the correlation is $\lambda \approx 2$.

A more general approach, known as Box-Cox transformation [4], can be used to identify the best transformation to achieve normality. The general form of the transformation is:

$$f(y) = \begin{cases} \frac{y^\lambda - 1}{\lambda} & \text{for } \lambda \neq 0 \\ \log(y) & \text{for } \lambda = 0. \end{cases}$$

Special cases are:

- $\lambda = -1$ then the transformation is the reciprocal,

- $\lambda = 0$ then the transformation is logarithmic,

- $\lambda = \frac{1}{2}$ then the transformation is a square root,

- $\lambda = 2$ then the transformation is a square.

The value of λ can be estimated either by maximum likelihood or by plotting the correlation coefficient of the QQ plot of the transformed data versus λ: λ is estimated as the value that maximises the correlation. Figure 5.6 shows such a plot for the data presented in Table B.2. The value of λ that maximises the correlation on the QQ plot is $\lambda \approx 2$ and corresponds to the square transformation.

Figure 5.7 shows the data, a histogram of the residuals and a QQ plot of the residuals following a Box-Cox transformation to the data in Table B.2. We can see that the histogram

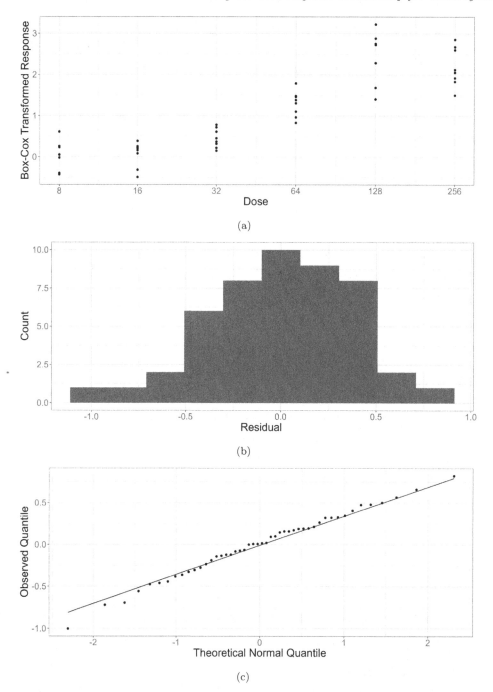

FIGURE 5.7

Box-Cox transformed response data (Table B.2) for a set of six doses in (a); histogram of
the residuals for the squared response in (b); and QQ-plot of the residuals for the squared
response in (c). There is no evidence that the data are not normally distributed.

is no longer skewed and the points of the QQ plot fall (roughly) along a straight line. The Shapiro-Wilk test p-value for the squared data is 0.9768, also indicating that there is no evidence to conclude that the data are non-normally distributed. The data can now be treated as normally distributed.

5.1.3 Impact of transformation of response on RP and concentration estimates

Transformation usually solves violations of both normality and homogeneity of variance assumptions. Even if we have apparent normality on the original scale, so transformation is not indicated, the transformed data may still be approximately normal.

It is important to understand that transformation of the response variable has no effect on the horizontal distance between two curves, and hence no effect on the underlying relative potency value. Similarly, transformation will have no impact on an interpolated concentration. However, transformation will affect the shape of the curves and the EC_{50}. See 5.8 where a log transformation of the response results in an exponential curve becoming a straight line.

Figure 5.9 illustrates the lack of effect on relative potency following transformation using an example. For the reference standard, at a dose of 102.4, we have response $R_S = 315.93$ and log response $\log R_S = 5.76$. For the test sample, at a dose of 409.6, we have response $R_T = 315.93$ and log response $\log R_T = 5.76$. That is, when $R_T = R_S$, we also have $\log R_T = \log R_S$. Therefore, a log transformation preserves the horizontal distance between the corresponding doses is unchanged and the relative potency in either case is 0.25.

Figure 5.10 illustrates the impact of a log transformation on the estimate of a concentration. The test sample has response $R_T = 334.63$ and log-transformed response $\log R_T = 5.81$. In either case, when interpolated from the standard curve plotted for both the (untransformed) response and log-transformed response, the interpolated concentration corresponds to a dose of 128.

It is important to note that when working with data, the estimate of the relative potency or the concentration based on transformed data will be different from the estimate of the relative potency for a log-transformed response. However, the true underlying relative potency itself is unaffected by a log transformation of the response. If an appropriate choice of transformation is made, the estimate of relative potency will be more accurate and precise than if the untransformed data were to be used.

5.1.4 Independence

Independence of replicates is best assessed by considering the laboratory procedures. Any time that the preparations involve shared experimental steps this can induce a dependency (correlation) among the responses. For example, in the preparation of a bioassay, when a new sample is taken from the batch to create each replicate this may result in independent (true) replicates. However, there still may be other factors that could cause dependencies. When

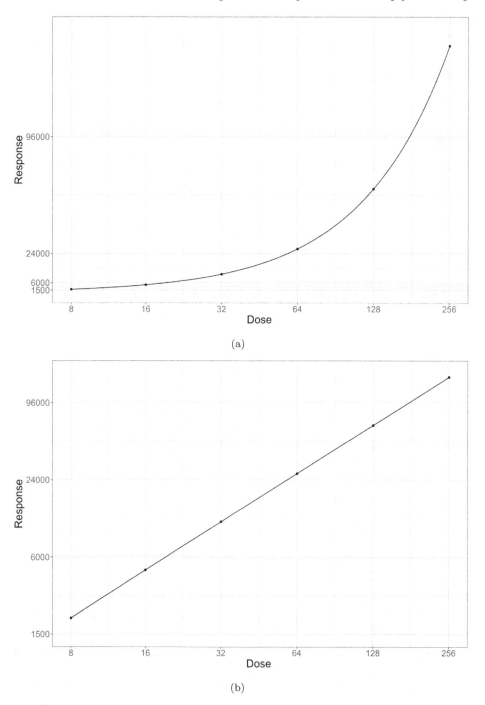

(a)

(b)

FIGURE 5.8

When the response is log-transformed, the shape of the dose-response relationship will change. On the original scale the relationship is exponential in (a); when the response is log-transformed the relationship becomes a straight line in (b).

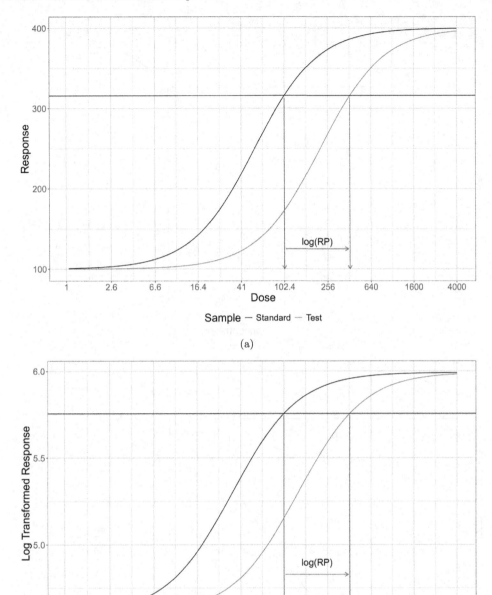

FIGURE 5.9

Dose-response curves for a reference test pair plotted with the untransformed response in (a) and log-transformed response in (b). The horizontal distance between the curves, and therefore the relative potency is $RP = 0.25$ in both cases.

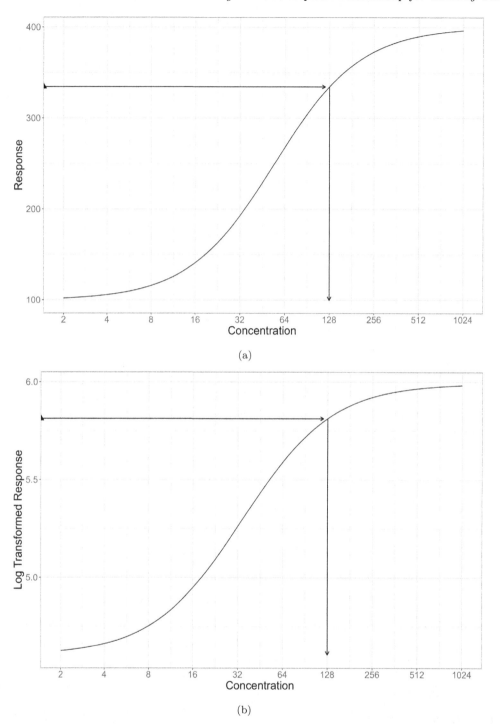

(a)

(b)

FIGURE 5.10

Estimation of a concentration when the response is untransformed in (a) and log-transformed in (b). In both cases, the estimate of the concentration is 128.

a group of replicates are prepared from a single sample from the batch, these will certainly not be independent and are sometimes referred to as 'pseudo' or 'technical' replicates.

From a statistical viewpoint, we rely on scientific advice as to whether the replicates are independent or not. If we do not have independence, then the error variance, or root-mean-squared error (which is needed for the calculation of confidence intervals) is likely to be underestimated if the individual values are included without careful consideration.

When dependencies exist, the simplest approach is to average the replicates before conducting any calculations. An alternative is to include the individual values but include the between-replicate variance in the model explicitly, ensuring that the variance calculations for confidence intervals are correctly carried out (see Chapter 6).

5.2 Functional forms for continuous dose-response relationships

The assumptions above apply to continuous response data regardless of the functional form of the dose-response relationship. We now discuss the functional form.

5.2.1 Straight-line relationship

Although a full S-shaped curve offers advantages in terms of assay range, there are sometimes practical reasons for focussing on a straight-line dose-response relationship. This is usually assumed to be representative of the central portion of a full S-shaped curve and is sometimes needed if either the bioassay system does not allow the upper and lower limits of response to be achieved (for reasons such as saturation) or where material is limited. For the special case of time-to-event data, the relationship is usually assumed to be a straight line and the response, time-to-event, is log-transformed.

5.2.1.1 Dose range for straight-line relationship

When establishing the dose range for a straight-line relationship, the central portion (or the 'linear part') of an S-shaped curve needs to be identified in terms of which dose levels are likely to produce responses that lie approximately on a straight line. In the first instance, we focus on the reference standard.

One approach is to characterise the full dose-response curve, then choose doses that correspond to a particular range of responses between the two asymptotes. A starting place could be to choose doses which produce responses in the middle 80% of the response range, that is, excluding the top 10% and the bottom 10%. Figure 5.11 illustrates this approach and highlights the linear portion of the curve. Another, more complex, approach for defining the linear portion is to define it as lying the portion of the curve between 'bend points' [55].

It is important to note that when choosing the set of doses for a straight-line relationship, the range of doses is narrow enough that a linear fit is reasonable. If the range of doses

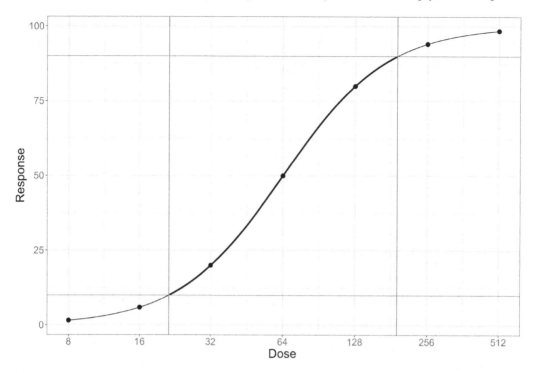

FIGURE 5.11

4PL dose-response relationship; horizontal lines have been plotted at 10% and 90% of the full range of responses. Vertical lines have been plotted at the corresponding doses; in this case there are three doses that fall in this central portion.

selected is too wide, then curvature in the dose-response relationship may be observed and a straight-line relationship may not fit the data well.

For test samples with relative potency values far from 1, the relationship may no longer fall in this linear portion (see Chapter 4 for further discussion). Therefore, test samples with potencies in the range of interest should be assessed as part of the bioassay development process to maximise the chance of obtaining a straight line. The wider the range of doses that can be used, the more precise the estimate of potency is likely to be, but the trade-off is that the dose-response for samples with potencies far from 1 will become curved.

5.2.1.2 Number of doses and spacing for straight-line relationship

Because the dose-response curve usually involves the concentrations on the log scale, dose groups should be chosen on a multiplicative scale. This results in the log doses being evenly spaced; it is also convenient in terms of making up the series in the laboratory. If the slope ratio approach is going to be used to calculate potency, this assumes a straight-line relationship with respect to the dose, and doses should be chosen on an additive (rather than multiplicative scale). Although a straight line can be fitted to two dose groups, data as limited as this cannot be assessed for the fit of the model (see Chapter 7), so at least three dose groups are required.

To provide the possibility of dropping dose groups that are outside the linear range for a particular test sample (or on a particular occasion) it is useful to include more than three dose groups.

5.2.2   S-shaped relationship

If the whole dose-response relationship can be modelled without reducing it to the central portion, the bioassay is likely to have a wider range of applicability. As discussed in Chapter 4, S-shaped relationships are usually described using either the four-parameter logistic (4PL) or the five-parameter logistic (5PL) relationships. Which is more appropriate for the bioassay can be assessed initially by eye. If there is a lack of (rotational) symmetry in the curve, then the 5PL may be required. Otherwise, the 4PL is preferable.

When choosing the doses for a S-shaped relationship, it is important first to identify if a transformation is required. The optimal set of doses will also depend on the response transformation. Figure 5.12 shows the an example of the impact of a log transformation on a 4PL dose-response relationship. In (a), the full relationship is represented by the 6 doses. However, in (b), following a log transformation there is no information on the lower asymptote for the same 6 doses.

5.2.2.1   Dose range, spacing and number of dose groups for S-shaped relationship

Again, because the dose-response curve usually involves the concentrations on the log scale, dose groups should be chosen on a multiplicative scale. The optimal set of doses will provide the maximum information about (a) the similarity of the reference standard curve and the test sample curve and (b) the relative potency estimate. Because the reference standard is used in every assay, and the test samples are expected generally to lie relatively close to the reference standard, it makes sense to choose the doses based on information about the reference standard.

The EC_{50}, or C parameter in the 4PL case, is dependent on the central portion. Data outside the central portion contribute to the estimation of the asymptotes. The slope parameter is dependent on all data (i.e., both the central portion and along plateaus). If the relationship truly has a 4PL curve, then observing data along one plateau coupled with sufficient data beyond the mid-point can still result in a reasonable estimate of both asymptotes. However, if the dose-response relationship is not symmetric (e.g., 5PL) this does not hold.

For the 4PL dose-response relationship at least five dose groups are required and at least six dose groups are required for the 5PL. A rule of thumb is to aim for at least 4 dose groups in the central 80% of the response range and a further 2 dose groups in each of the top and bottom 10% (towards at least one of the asymptotes). Similar considerations apply for the 5PL, with the exception that because the curve is not symmetric, data are needed at both ends of the response range to allow estimation of the asymptotes. These considerations apply to the standard curve for interpolation-type bioassays as well.

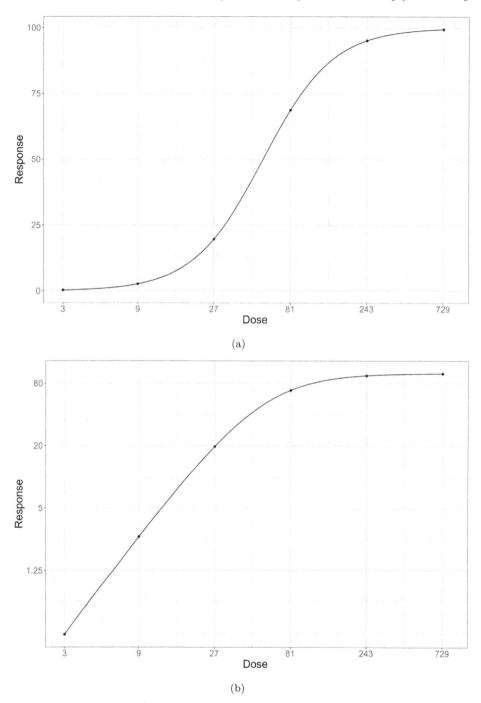

(a)

(b)

FIGURE 5.12

When choosing the doses for a S-shaped relationship, it is important first to identify if a transformation is required. In (a), the full relationship is represented by the 6 doses. However, in (b), following a log transformation there is no information on the lower asymptote for the same 6 doses.

5.3 Binary data

For assays with a binary response, the response for each individual subject will be either a 1 or a 0. For dose group i, $i = 1, \ldots, d$, with n_i subjects, the number of responses that take on a value of 1, Y_i, is assumed to have a binomial distribution. That is:

$$Y_i \sim \text{Bin } (n_i, p_i).$$

For dose group i, the mean, or the expected number of observations to take on a value of 1, is:

$$\mu_i = n_i \times p_i$$

with variance:

$$\sigma_i^2 = n_i \times p_i \times (1 - p_i).$$

Therefore, the variance is not constant across response values. When $p_i = 0$ or $p_i = 1$ we can consider these as sure events and the variance will be zero. For binary data, the greatest uncertainty occurs if $p_i = 0.5$.

To provide an example, consider an experiment where the efficacy of a toxin is being investigated by administering that toxin to animals at various dose levels. At a zero dose, it may be reasonable to expect that all the animals will survive the study (all responses will take on a value of 0). Then $p_i = 0$ and the variance will also be zero. A similar scenario may occur at a very high dose if all animals are expected to die.

The models we have discussed for continuous data assume a normal distribution with constant variance. However, these assumptions do not apply for binary data and accounting for this is necessary when selecting a dose-response model. For binary data, the model is always either a logit or a probit model as described in Chapter 4 and there is very little difference between these in practice. A test can be performed to assess if the pattern of variation is consistent with a binomial distribution.

As for the continuous data case, the logit model assumes independence among the observations. If multiple animals are located in the same cage then this assumption may not hold. When observations are dependent then alternative models such as generalised estimating equations [34, 70] or generalised linear mixed models [28] can be used to account for the correlation.

The considerations required for making the choice of doses for the logit and probit models are similar to those for S-shaped curves for continuous data. In general, it is advisable to have at least 2 non-extreme results, that is with proportions nearer to 0.5. However, a more conservative rule of thumb is to aim for at least 4 dose groups in the central 80% of the response range and a further 2 dose groups in each of the top and bottom 10% (towards each asymptote). The model cannot be fitted if there are fewer than 2 groups with responses between 0 and 1.

5.4   Chapter summary

- The choice of dose-response relationship, including both the shape of the dose-response relationship and the scatter of responses around the fitted curve, is considered.

- For inferences to be made about the potency of a sample, including the calculation of a confidence interval, for continuous data, the response data must (at least approximately) follow a normal distribution, and have homogeneous variance across the response range. For binary data, the responses should follow a binomial distribution. For both types of response, the observations within and between doses must be independent.

- Methods are presented for determining whether or not the assumptions are justified, including QQ plots and the Shapiro-Wilk test. Where the assumptions are not justified, potential solutions are provided, including the management of pseudo replicates, where there are dependencies, and selection of a transformation or weights to the response data. Using a Box-Cox transformation to inform the choice of transformation is explained and illustrated.

- The impact of extreme observations, or outliers, on the assessment of the dose-response relationship, is discussed. The number of doses to be used, and their spacing, are discussed for the different dose-response relationships.

6

Bioassay data analysis

Once the mathematical model for the dose-response relationship is chosen, the values of the parameters must be estimated from the data. The method to be used for the calculation of the parameter estimates depends on the form of the response: continuous, binary, or time-to-event. It also depends on the design of the bioassay. In general, we wish to choose the parameters estimates that result in the line most closely fitting the observed data points.

One approach to parameter estimation commonly used when the response is continuous is to minimise the residual sum of squares (RSS). The RSS is the sum of the squared distances between the individual data points and the fitted curve and characterises how well this curve is fitting the data. The resulting estimates are called the least squares estimates. For time-to-event and binary data, maximum likelihood estimation is typically used. Maximum likelihood estimation is beyond the scope of this book and instead we refer to [14, 32]. In the special case of continuous data, with normally distributed errors, maximum likelihood estimation and least squares estimation are identical.

An explicit set of formulae for the parameter estimates is only available for the case of a straight-line relationship. In the other cases we cover, iterative numerical procedures are required to minimise the residual sum of squares (or to maximise the likelihood). We provide an overview of the use of iterative methods to minimise the residual sum of squares for the 4PL relationship below. In general, the calculation of estimates of the parameters including the RP, as well as their confidence intervals, require the use of statistical software. We provide the formulae and background in this chapter to aid in understanding.

At initialisation an iterative numerical method begins with a rough guess of the values of the parameters (for example A, B, C and D for the 4PL). These 'starting values' may be quite far from optimal. The RSS resulting from the curve defined by the starting values is calculated and then the parameter estimates are adjusted. If the residual sum of squares decreases after an update to the parameter estimates is made, then these new values are accepted and the process is repeated. If the residual sum of squares does not decrease, then an alternative update to the parameter estimates may be proposed. There are many different parameter estimation algorithms, or ways of choosing how updates to the parameter estimates are made. These iterative processes continue until it is impossible to find a change that decreases the residual sum of squares, in which case the software concludes that it has found the optimal set of parameters that provide the best possible fit to the data. This finding of a solution is sometimes referred to as model convergence.

DOI: 10.1201/9781003449195-6

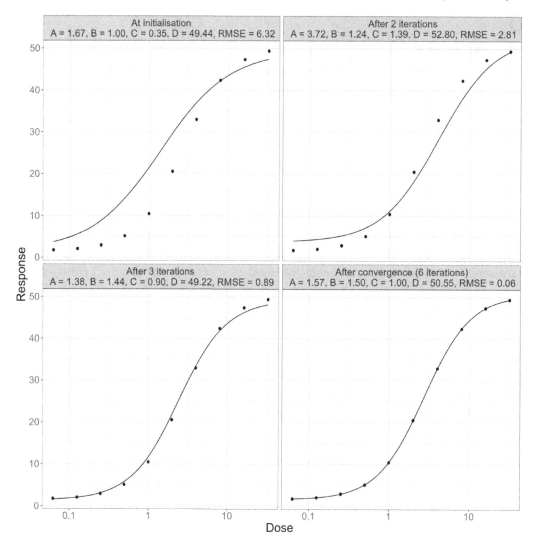

FIGURE 6.1

Illustration of an iterative numerical procedure for fitting a 4PL curve. With each subsequent iteration, the curve provides a closer fit to the data. The root mean squared error (RMSE) is a function of the RSS and characterises how closely the model fits the data; the smaller the RMSE, the closer the fit.

Figure 6.1 shows an an example of the iterative estimation procedure for a 4PL model. At initialisation, the fitted curve lies quite far from the data points. As the process continues, with each update to the parameters, the curve provides a closer fit to the data points. After 6 iterations the process reaches convergence and the model provides a very close fit to the data.

In some situations the results may be unexpected, in the sense that the fitted model does not appear to match the data points. For example, it is possible that the 'best' fit determined by the software is still very poor. This can happen if we try to fit bioassay data to an inappropriate model. If the data are asymmetric but we try to fit a 4PL, the fit cannot be close to the data everywhere, since the 4PL is inherently symmetric. However, in other cases, the procedures can sometimes give a suboptimal result. We include some examples in this chapter.

We start with the simple case where we assume that the only factors relevant to estimation of concentration or potency are the dose and the sample. This may be not ideal in the presence of row and column effects, or if data from several plates are analysed together. If these effects are present then it may be possible to control for them within the statistical analysis, to achieve a more precise estimate of the potency. This more sophisticated analysis method is presented in Section 6.4.

6.1 Fitting the dose-response relationship to continuous data

We assume that we have a set of n independent observations labelled $i = 1, \ldots, n$ consisting of the dose-response pair (dose_i, y_i). The individual response values, transformed if necessary, (see Chapter 5) are assumed to be normally distributed with mean given by the dose-response relationship and a variance which is constant for all dose groups.

6.1.1 Straight-line relationship

For the straight-line relationship, the ith response value (transformed if necessary), y_i, can be expressed as:

$$y_i = \beta_0 + \beta_1 \times \log(\text{dose}_i) + \epsilon_i, \quad i = 1, \ldots, n$$

where:

β_0, β_1 are the intercept and slope parameters,

ϵ_i is a normally distributed error term with mean 0 and variance σ^2.

The values of β_0, β_1, and σ^2 are unknown and must be estimated from the data. The explicit formulae for β_0 and β_1 can be found in Appendix A. For the pair of estimates, $\hat{\beta}_0$, $\hat{\beta}_1$, the fitted value for a given dose is:

$$\hat{y}_i = \hat{\beta}_0 + \hat{\beta}_1 \times \log(\text{dose}_i).$$

The residual for the ith observation, e_i, is:

$$\begin{aligned} e_i &= y_i - \hat{y}_i \\ &= y_i - \left[\hat{\beta}_0 + \hat{\beta}_1 \times \log(\text{dose}_i) \right]. \end{aligned} \tag{6.1}$$

The estimate of the standard deviation, $\hat{\sigma}$, is given by the root-mean-squared error (RMSE):

$$\hat{\sigma} = \sqrt{\frac{\sum_{i=1}^{n} (y_i - \hat{y}_i)^2}{n - p}}, \tag{6.2}$$

TABLE 6.1

Summary of parameter estimates for the straight-line relationship fitted to the reference standard data in Table B.3.

	Estimate	Std. error	95% CI
Intercept, $\hat{\beta}_{0S}$	0.944	0.037	(0.864, 1.024)
Slope $\hat{\beta}_{1S}$	0.224	0.005	(0.213, 0.236)
RMSE, $\hat{\sigma}_S$	0.068		

where p is the number of parameters; for a straight-line relationship fitted to a single sample, $p = 2$ (β_0, β_1). The numerator of the RMSE is the sum of the squared residuals, while the denominator is the corresponding degrees of freedom. This estimate of the standard deviation is used to estimate the confidence intervals for the parameters.

Note that these 'model-based' residuals are with respect to the straight line and not to the mean of the dose group. The latter residuals were discussed in Chapter 5 and were used to assess normality of the responses per group. The 'model-based' residuals are fundamental to the calculation of the estimates for the dose-response relationship and their confidence intervals, and they can be used to assess extreme data points (see Section 6.5) as well as other departures from the relationship (see Chapter 7).

Figure 6.2 shows the straight-line dose-response relationship, along with the corresponding residuals, for the reference standard data of the example data set in Table B.3. The parameter estimates, along with their confidence intervals, are summarised in Table 6.1.

6.1.1.1   Relative potency estimation

For the estimation of relative potency, a parallel straight-lines model must be fitted to the data for the standard and test samples. In this case we assume that the total number of observations $n = n_S + n_T$, where n_S is the number of data points for the standard and n_T is the number of data points for the test sample. The parallel straight-line relationship is defined as:

$$y_i = \beta_{0S} + [\beta_1 \times \log(\text{dose}_i)] + [\Delta_{\beta_0} \times x_i] + \epsilon_i, \tag{6.3}$$

where y_i is the observed response for the ith observation, with corresponding dose given by dose_i, and x_i is an indicator variable that takes on a value of 0 if the sample corresponding to the ith observation is the reference standard and 1 if the test sample. For this parameterisation of the parallel straight-line relationship, Δ_{β_0} is the intercept difference (test - reference standard) and so the log RP is given by:

$$\log \text{RP} = \frac{\Delta_{\beta_0}}{\beta_1}.$$

The triplet of values, $\hat{\beta}_{0S}$, $\hat{\Delta}_{\beta_0}$, $\hat{\beta}_1$, for the intercept, intercept difference, and slope that result in the minimum possible sum of squared residuals across all data, is the set of least squares estimates for the parallel lines. The estimate of the standard deviation, $\hat{\sigma}$, is again

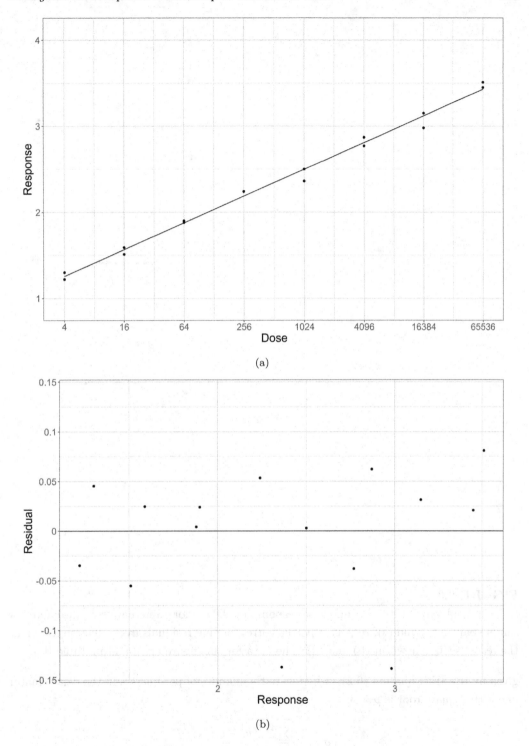

(a)

(b)

FIGURE 6.2
Straight-line dose-response model fit to the reference standard data in Table B.3 in (a) and
the corresponding residuals in (b).

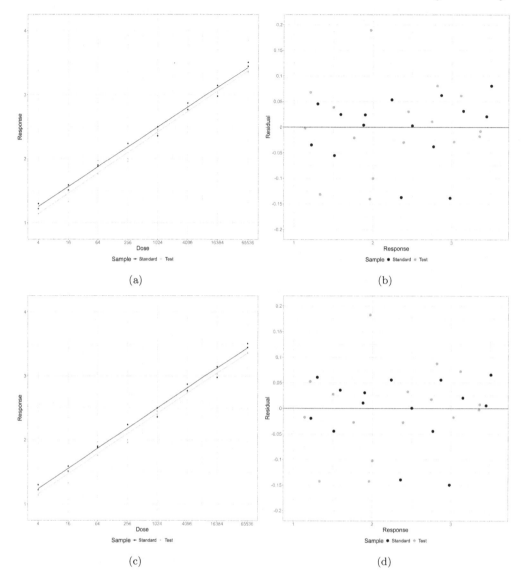

FIGURE 6.3

Separate and parallel straight-line dose-response model fits for a reference standard and test sample pair in (a) and (c), respectively; the corresponding residuals from each model versus the response are shown in (b) and (d). The data for this figure are found in Table B.3.

given by RMSE (Equation (6.2)), where here the number of parameters in the denominator under the square root is $p = 3$:

$$\hat{\sigma} = \sqrt{\frac{\sum_{i=1}^{n} \left(y_i - \hat{y}_i\right)^2}{n - 3}}.$$

This value is used to estimate the confidence intervals for the parameters and RP.

Figure 6.3 shows the separate and parallel fits to data for the reference standard and test sample data in Table B.3, along with the residuals. Table 6.2 summarises the parameter

TABLE 6.2

Summary of parameter estimates for the straight-line relationship fit to the reference standard and test sample separately, as well as the parallel model for the data in Table B.3.

	Estimate	Std. error	95% CI
Separate-lines (unconstrained)			
Reference standard			
Intercept, $\hat{\beta}_{0S}$	0.944	0.037	(0.864, 1.024)
Slope, $\hat{\beta}_{1S}$	0.224	0.005	(0.213, 0.236)
RMSE, $\hat{\sigma}_S$	0.068		
Test sample			
Intercept, $\hat{\beta}_{0T}$	0.823	0.047	(0.721, 0.924)
Slope, $\hat{\beta}_{1T}$	0.230	0.007	(0.216, 0.245)
RMSE, $\hat{\sigma}_T$	0.086		
Parallel-lines (constrained)			
Intercept (reference standard), $\hat{\beta}_{0S}$	0.924	0.033	(0.857, 0.991)
Intercept difference, $\hat{\Delta}_{\beta_0}$	−0.082	0.027	(−0.137, −0.026)
Slope (common), $\hat{\beta}_1$	0.227	0.004	(0.219, 0.236)
RMSE, $\hat{\sigma}$	0.077		
RP	0.697		(0.546, 0.890)*

*Confidence interval for the relative potency calculated using Fieller's theorem.

estimates for the separate and parallel fits. The separate-lines model achieves the closest possible straight-line fit for each sample. When the slopes are constrained to be parallel, the model is fitted with the same slope for both samples. As a result, this model does not fit the data for either sample as closely when compared to fitting the relationship for each sample separately and the residuals are larger. However, fitting a parallel model is required when estimating a relative potency.

A confidence interval for $\log \mathrm{RP} = \frac{\Delta_{\beta_0}}{\beta_1}$, can be obtained by Fieller's Theorem [18]. However, it can fail in certain scenarios, and we present the methodology to demonstrate this. The calculation of confidence intervals for model parameters (including the RP) is typically handled by statistical software.

The Fieller's Theorem 95% confidence limits for the RP are given by:

$$\frac{1}{1-g} \cdot \left[\frac{\hat{\Delta}_{\beta_0}}{\hat{\beta}_1} - \frac{g \cdot \nu_{12}}{\nu_{22}} - \frac{t_{0.975,n-3} \cdot \hat{\sigma}}{\hat{\beta}_1} \cdot \sqrt{\nu_{11} - 2 \cdot \frac{\hat{\Delta}_{\beta_0}}{\hat{\beta}_1} \cdot \nu_{12} + \frac{\hat{\Delta}_{\beta_0}^2}{\hat{\beta}_1^2} \cdot \nu_{22} - g \cdot \left(\nu_{11} - \frac{\nu_{12}^2}{\nu_{22}} \right)} \right]$$

and

$$\frac{1}{1-g} \cdot \left[\frac{\widehat{\Delta}_{\beta_0}}{\hat{\beta}_1} - \frac{g \cdot \nu_{12}}{\nu_{22}} + \frac{t_{0.975,n-3} \cdot \hat{\sigma}}{\hat{\beta}_1} \cdot \sqrt{\nu_{11} - 2 \cdot \frac{\widehat{\Delta}_{\beta_0}}{\hat{\beta}_1} \cdot \nu_{12} + \frac{\widehat{\Delta}_{\beta_0}^2}{\hat{\beta}_1^2} \cdot \nu_{22} - g \cdot \left(\nu_{11} - \frac{\nu_{12}^2}{\nu_{22}} \right)} \right],$$

where:

$\widehat{\Delta}_{\beta_0}$ is the estimate of Δ_{β_0},

$\hat{\beta}_1$ is the estimate of β_1,

$\hat{\sigma}$ is the root-mean-squared-error from the model,

$\nu_{11}\hat{\sigma}^2$ is the estimate of the variance of β_1,

$\nu_{12}\hat{\sigma}^2$ is the estimate of the covariance of β_1 and Δ_{β_0},

$\nu_{22}\hat{\sigma}^2$ is the estimate of the variance of Δ_{β_0},

$t_{0.975,n-3}$ is the 97.5th percentile of the t distribution with $n-3$ degrees of freedom and

g is a constant defined as:

$$g = \frac{t_{0.975,n-3} \cdot \hat{\sigma}^2 \cdot \nu_{22}}{\hat{\beta}_1^2}.$$

There are three situations in which this interval is not appropriate and alternative methods need to be used:

1. If the quantity under the square root is negative, then the limits are not defined;

2. If $g=1$, then the limits are infinite;

3. If $g>1$, then the confidence interval consists of two disjoint intervals.

The delta method [46] provides an alternative.

6.1.2 4PL relationship

When the dose-response relationship has the 4PL shape, we again assume that the individual data values are normally distributed with mean value given by the dose-response relationship and a variance which is constant for all dose groups. In the case of a 4PL relationship, the ith response value (transformed if necessary), y_i, is expressed as:

$$y_i = D + \frac{A-D}{1 + e^{B(\log(\text{dose}_i)-C)}} + \epsilon_i, \quad i = 1, \ldots, n,$$

where:

A, B, C, D are the 4PL parameters,

ϵ_i is a normally distributed error term with mean 0 and variance σ^2.

TABLE 6.3

Summary of parameter estimates for the 4PL relationship fitted to the reference standard data in Table B.4.

	Estimate	Std. error	95% CI
Left asymptote, $\hat{A}_S$	0.991	0.018	(0.953, 1.029)
Slope parameter, $\hat{B}_S$	1.455	0.042	(1.369, 1.541)
log EC$_{50}$, $\hat{C}_S$	4.011	0.020	(3.969, 4.053)
Right asymptote, $\hat{D}_S$	4.008	0.021	(3.964, 4.052)
RMSE, $\hat{\sigma}_S$	0.041		

The values of the 4PL parameters, A, B, C, D, as well as the standard deviation, σ, must be estimated from the data. Again the usual approach is to use least squares (or equivalently maximum likelihood) estimation. That is, where the estimates, $\hat{A}$, $\hat{B}$, $\hat{C}$, $\hat{D}$, are chosen such that they result in the minimum possible sum of squared residuals. The estimate of the standard deviation, $\hat{\sigma}$, is given by the RMSE (Equation (6.2)), where now $p = 4$. Again, this value is used to estimate the confidence intervals for the parameters. The confidence intervals for the parameters are typically given by software.

Figure 6.4 shows the 4PL dose-response model fit to the reference standard data in Table B.4 along with the residuals. The parameter estimates, along with their confidence intervals, are summarised in Table 6.3.

Points to note

For the 4PL curve, certain data configurations can lead to a situation in which there is no finite solution to the model fitting process. For example, with sparse data in the central portion of the response range, an estimate of the slope parameter may be infinite. Figure 6.5 provides an example of this.

With bioassay data like these, the residual sum of squares will always become smaller as B is increased. Mathematically, the best fit (the solid line) has infinite B, which means the model curve is flat all the way up to the single dose corresponding to an intermediate response, where it abruptly jumps to the other asymptote.

The iterative algorithms used to estimate the model parameters can never reach infinity. Instead, software will often report whatever value of B was reached on the final iteration and may be a large value such as 10 or 20. This will give a steep model curve, like the light grey curve in the plot, but without an abrupt jump. There may be no obvious sign that anything is wrong, but using a value of 10, or a value of 20, for B will not be correct. When this occurs, it is likely that the confidence interval for B will be extremely wide.

To avoid this problem, it is important to examine the software output for warning messages, and consider using suitability criteria which reject fits with very large values of B. If this happens repeatedly, it may be necessary to decrease the spacing between doses to get more data on the central portion of the curve.

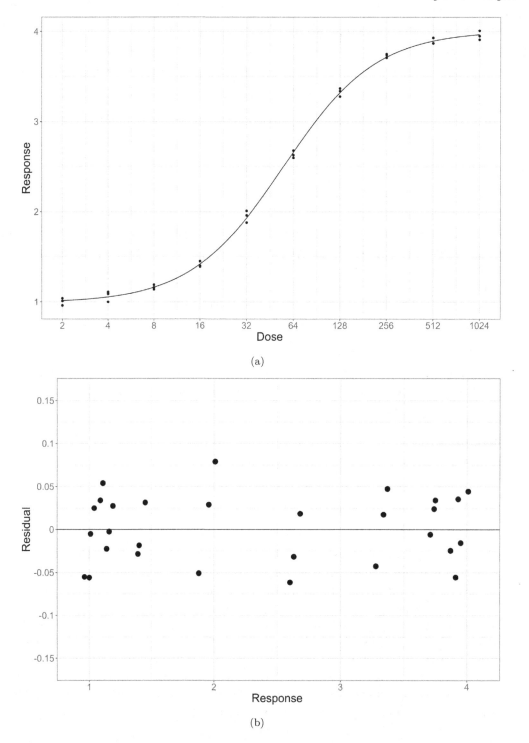

(a)

(b)

FIGURE 6.4

4PL fit to the reference standard data in Table B.4 in (a) and the corresponding residuals in (b).

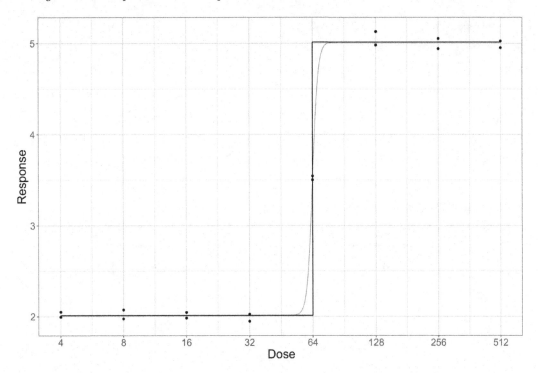

FIGURE 6.5

Dose-response relationship with sparse data in the central portion of the curve. The best fit line has infinite slope (shown in black); bioassay software may return a suboptimal fit for the 4PL curve (shown in grey).

6.1.2.1 Relative potency estimation

For the estimation of RP, parallel models must be fitted to the data for the standard and test samples. Again we assume a total of n data points comprised of n_S data points for the standard and n_T data points for the test sample, with parallel model:

$$y_i = D + \frac{A - D}{1 + e^{B[\log(\text{dose}_i) - C_S + (\Delta_C \times x_i)]}} + \epsilon_i, \tag{6.4}$$

where $\Delta_C = C_S - C_T$ and corresponds to the log RP and x_i is an indicator variable as defined for Equation (6.3). The set of five values, $\hat{A}$, $\hat{B}$, $\hat{C}_S$, $\hat{\Delta}_C$, $\hat{D}$ that result in the minimum possible sum of squared residuals for both standard and test sample data combined, is the set of least squares estimates for the parallel curves. The estimate of the standard deviation, $\hat{\sigma}$, is given by the RMSE (Equation (6.2)), where $p = 5$.

It is important to note that different software packages use different formulations for the 4PL dose-response relationship, so the meaning of the four parameters varies from package to package. However, the different formulations have no effect on the RP.

Figure 6.6 shows the separate and parallel fits to data for the reference standard and test sample data in Table B.4, along with the residuals. Table 6.4 summarises the parameter estimates for the separate and parallel fits. For the parallel model fit, both asymptotes and the slope have been constrained to be the same for both samples.

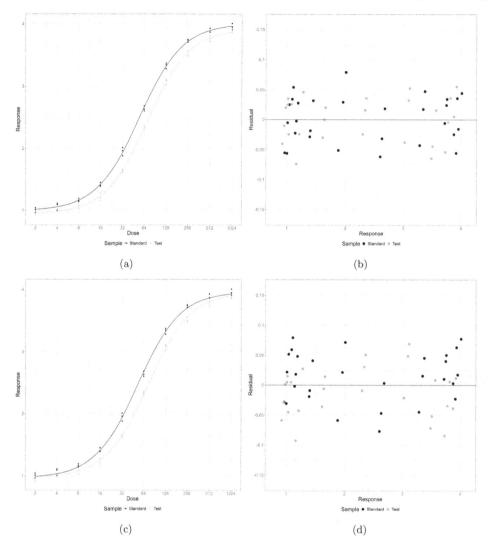

FIGURE 6.6
Separate dose-response model fit to two samples in (a) with corresponding residuals in (b); parallel dose-response model fit to the same data in (c) with corresponding residuals in (d). In general, the residuals for the parallel fit are larger. The data for this example are in Table B.4.

6.1.3　5PL relationship

For a 5PL dose-response relationship, we again assume that the individual data values are normally distributed with mean value given by the dose-response relationship and a variance which is constant for all dose groups. The ith response value (transformed if necessary), y_i, is expressed as:

$$y_i = D + \frac{A - D}{\left(1 + e^{B(\log(\text{dose}_i) - C)}\right)^E} + \epsilon_i, \;\; i = 1, \ldots, n$$

where:

TABLE 6.4

Summary of the parameter estimates for the 4PL relationship fitted separately to the reference standard and test sample, and the parallel 4PL relationship for the data in Table B.4.

	Estimate	Std. error	95% CI
Separate-lines (unconstrained)			
Reference standard			
Left asymptote, $\hat{A}_S$	0.991	0.018	(0.953, 1.029)
Slope parameter, $\hat{B}_S$	1.455	0.042	(1.369, 1.541)
log EC$_{50}$, $\hat{C}_S$	4.011	0.020	(3.969, 4.053)
Right asymptote, $\hat{D}_S$	4.008	0.021	(3.964, 4.052)
RMSE, $\hat{\sigma}_S$	0.041		
Test sample			
Left asymptote, $\hat{A}_T$	0.946	0.015	(0.915, 0.978)
Slope parameter, $\hat{B}_T$	1.506	0.041	(1.422, 1.590)
log EC$_{50}$, $\hat{C}_T$	4.255	0.019	(4.216, 4.294)
Right asymptote, $\hat{D}_T$	3.917	0.021	(3.873, 3.961)
RMSE, $\hat{\sigma}_T$	0.038		
Parallel-lines (constrained)			
Left asymptote (common), $\hat{A}$	0.964	0.014	(0.937, 0.991)
Slope parameter (common), $\hat{B}$	1.464	0.033	(1.399, 1.529)
log EC$_{50}$ (reference standard), $\hat{C}_S$	3.968	0.019	(3.930, 4.005)
log EC$_{50}$ difference, $\hat{\Delta}_C$	−0.336	0.020	(−0.376, −0.295)
Right asymptote (common), $\hat{D}$	3.972	0.017	(3.937, 4.006)
RMSE, $\hat{\sigma}$	0.044		
RP	0.715		(0.686, 0.745)

A, B, C, D, E are the 5PL parameters,

ϵ_i is a normally distributed error term with mean 0 and variance σ^2.

The values of A, B, C, D, E and σ must be estimated from the data in a manner analogous to the 4PL. We do not show the parallel model but it is the same as for the 4PL (Equation (6.4)), but where the denominator is raised to the exponent of E.

Figure 6.7 shows separate 5PL fits to the reference standard and test sample as well as the resulting residuals for the example data in Table B.5. It also shows the parallel 5PL model fit to the same data as well as the residuals for the parallel model fit. The estimates of the model parameters for both the separate and parallel fits are summarised in Table 6.5.

TABLE 6.5

Summary of parameter estimates for the 5PL relationship fitted separately to the reference standard and test sample, and the parallel 5PL relationship for the data in Table B.5.

	Estimate	Std. error	95% CI
Separate-lines (constrained)			
Reference Standard			
Left asymptote, $\hat{A}_S$	141.384	2.521	(136.193, 146.575)
Slope parameter, $\hat{B}_S$	−1.346	0.130	(−1.615, −1.078)
$\hat{C}_S$	6.003	0.213	(5.565, 6.441)
Right asymptote, $\hat{D}_S$	477.914	3.507	(470.692, 485.137)
Asymmetry parameter, $\hat{E}_S$	0.627	–	(0.405, 0.970)*
RMSE, $\hat{\sigma}_S$	5.542		
Test Sample			
Left asymptote, $\hat{A}_T$	152.427	2.511	(147.256, 157.598)
Slope parameter, $\hat{B}_T$	−1.853	0.241	(−2.350, −1.355)
$\hat{C}_T$	6.797	0.139	(6.510, 7.084)
Right asymptote, $\hat{D}_T$	498.512	4.714	(488.803, 508.221)
Asymmetry parameter, $\hat{E}_T$	0.313	–	(0.204, 0.482)*
RMSE, $\hat{\sigma}_T$	5.991		
Parallel-lines (unconstrained)			
Left asymptote (common), $\hat{A}$	146.737	2.065	(142.597, 150.877)
Slope parameter (common), $\hat{B}$	−1.562	0.142	(−1.846, −1.278)
Reference standard, $\hat{C}_S$	6.276	0.139	(5.998, 6.554)
Difference, $\hat{\Delta}_C$	−0.346	0.039	(−0.425, −0.268)
Right asymptote (common), $\hat{D}$	488.620	3.389	(481.825, 495.415)
Asymmetry parameter (common), $\hat{E}$	0.434	–	(0.311, 0.606)*
RMSE, $\hat{\sigma}$	6.672		
RP		0.707	(0.654, 0.765)

*The confidence interval for the asymmetry parameter E is calculated on the log scale and its standard error has been omitted.

Points to note

The 5PL model presents a significant challenge to the model fitting process when compared to the 4PL model. This occurs because in some scenarios many combinations of the 5 parameters can lead to very similar residual sums of squares. Therefore, there are many possible model fits that lead to curves that are similarly close to the observed data points. It is noted in [24] that it is possible to make large changes to the parameters such that the curves differ substantially in areas of no data, but yet still offer a very similar fit to the data points.

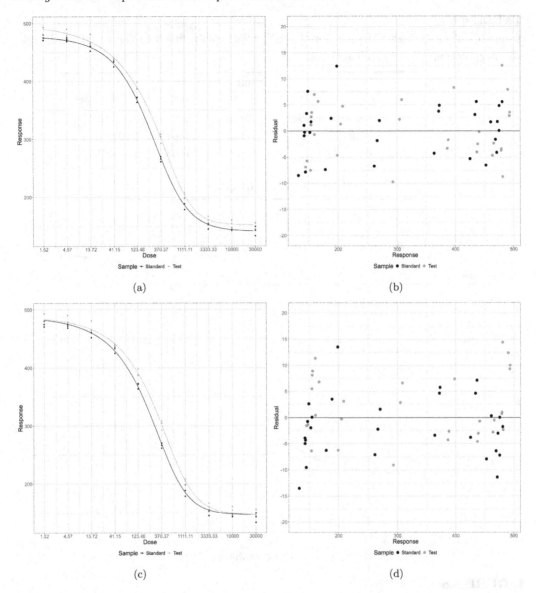

(a) (b)

(c) (d)

FIGURE 6.7
Separate 5PL fits to a reference standard and test sample in (a) and their residuals in (b).
Parallel 5PL fit to the same data in (c) with the residuals to the parallel model fit in (d).
The data for this example are presented in Table B.5.

6.1.4 Concentration estimation: conventional interpolation

In an interpolation-type assay, the conventional approach to estimating the concentration
of the test sample is to read the concentrations for the test sample dilutions from the
standard curve. In practice this involves rearranging the equation for the concentration
(dose) response relationship; this then allows for calculation of the concentration required
to yield a given response. These concentrations are then corrected for the dilution factor
and combined using the geometric mean. In this section we demonstrate interpolation from

TABLE 6.6

Summary of parameter estimates for the 4PL relationship fitted to the standard data for an interpolation bioassay in Table B.6.

	Estimate	Std. error	95% CI
Left asymptote, $\hat{A}_S$	2.518	0.029	(2.448, 2.588)
Slope parameter, $\hat{B}_S$	2.031	0.049	(1.911, 2.151)
log EC$_{50}$, $\hat{C}_S$	7.133	0.013	(7.101, 7.165)
Right asymptote, $\hat{D}_S$	9.484	0.035	(9.398, 9.570)
RMSE, $\hat{\sigma}_S$	0.0139		

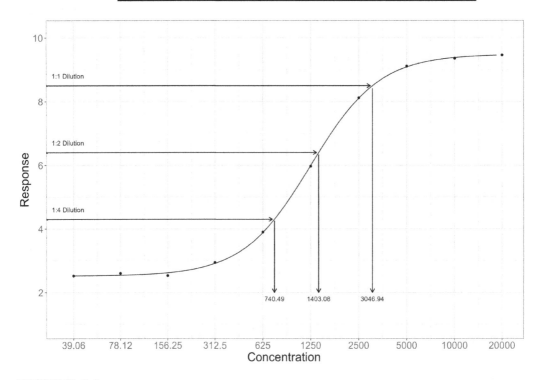

FIGURE 6.8

Interpolation of the concentration at three dilutions of a test sample. For different dilutions of the test sample, the corresponding concentration of the standard that resulted in the same response is recorded. The data for this example are in Table B.6.

a 4PL model for the standard curve. However, other relationships for the standard curve also apply.

Figure 6.8 shows the 4PL fit to the standard data presented in the example data set in Table B.6 of Appendix B. The 4PL model parameter estimates for the standard curve are summarised in Table 6.6. The interpolated concentrations for each dilution and the final estimate of concentration combined across the three dilutions are shown in Table 6.7.

TABLE 6.7

Calculation of the estimated concentration of a test sample for the data presented in Table B.6.

Dilution	Response	Concentration	Corrected concentration
1:1 Dilution	8.5	3046.94	$1 \times 3046.94 = 3046.94$
1:2 Dilution	6.4	1403.08	$2 \times 1403.08 = 2806.15$
1:4 Dilution	4.3	740.49	$4 \times 740.49 = 2961.98$
	Geometric Mean Result		2936.65

There is an implicit assumption, in the correction for the dilution, that the test sample dilution-response curve is parallel to the standard curve. That is, each corrected concentration is an estimate of the same quantity. We develop this further in the next section.

6.1.4.1   Concentration estimation: parallel curves

An improved method for estimating concentration is to build on the underlying assumption that the test sample responses and their dilutions lie on a curve that is parallel to the standard curve. If there are enough dilutions for the test sample that a 4PL curve can be fitted to it, then a relative potency value can be estimated as above, and the concentration obtained by multiplying by the concentration on the standard curve that corresponds to a dilution of 1.

If there are too few dilutions of the test sample to allow a 4PL curve to be fitted to it, then the following approach can be taken [9]. First the standard curve 4PL model is estimated. Then a 4PL curve is fitted to the test sample with the asymptotes and the slope parameter fixed at the standard curve values, that is, the 4PL curve parallel to the standard curve which best fits the test sample data. The RP for the test sample can then be estimated as usual and the concentration can be calculated as usual for a RP bioassay.

This method has been shown to be superior to the conventional approach described in Section 6.1.4 in several ways [9]. It is more accurate and precise than the conventional approach. It is not confined to response values that lie between the standard curve asymptotes. As a result, it can provide a concentration estimate for a wider range of samples. Therefore, the need to repeat a test with a different dilution may be decreased. The concentration estimate can be accompanied by a confidence interval which can be used to assess whether it is acceptably precise.

Figure 6.9 shows how this approach can be used for the data presented in Table B.6 of Appendix B. In this case, the estimate of the concentration is 2888.94 with a 95% confidence interval of (2756.77, 3027.44). The estimate of the concentration using this method similar to that of the conventional analysis estimate of 2936.65 (Section 6.1.4).

6.1.5   Time-to-event data

Although time-to-event data are continuous, they need special treatment when the dose-response relationship is being fitted to the data. This is because the event in question may not yet have happened at the end of the study, so its timing is not known. An example is

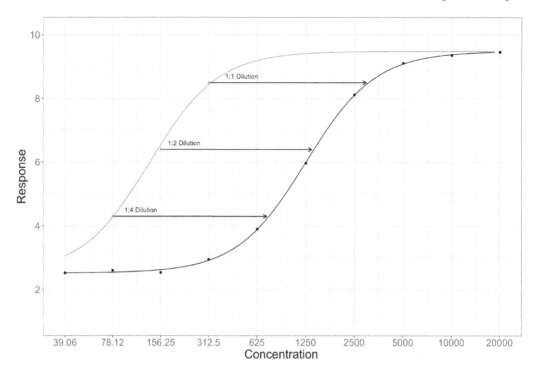

FIGURE 6.9

Interpolation of a concentration by fitting a 4PL relationship to the interpolated points with A, B, and D parameters fixed at those of the standard.

in a vaccine challenge study, where the event is death. If an animal is still alive at the end of the study, its time to death is known to be at least the duration of the study, but the exact value is unknown. This is known as a 'censored' observation.

Censored observations contribute information to the dose-relationship differently from data values that are exactly measured. A statistical technique known as 'survival analysis' (from clinical applications) or 'reliability analysis' (from engineering applications) is appropriate for data such as these [29, 32]. The technique is based on maximum likelihood. We mention these methods for completeness, and because they have been shown to improve the efficiency of the analysis. That is, the number of animals can be reduced whilst maintaining the same precision for the final estimate [69]. However, the data analysis requires statistical software and expertise and is not generally implemented in bioassay packages. Therefore, we do not provide details.

6.2   Fitting the dose-response relationship to binary data

When the response for an individual subject is binary an event occurs or it does not occur. In this case, each subject has a probability of the event occurring that is related to the dose given.

TABLE 6.8

Parameter estimates for the parallel logit dose-response relationship for the data in Table B.7.

	Estimate	Std. error	95% CI
Intercept (reference), $\hat{\beta}_{0S}$	2.013	0.299	(1.427, 2.599)
Slope parameter (common), $\hat{\beta}_1$	1.964	0.185	(1.601, 2.327)
Intercept difference, $\hat{\Delta}_{\beta_0}$	0.671	0.295	(0.093, 1.248)
RP	1.407		(1.050, 1.893)*

*Confidence interval for RP obtained using Fieller's theorem.

The response for each subject is coded as 0 or 1, where 1 corresponds to the case that the event has occurred. Let y_i be the response for the ith subject, $i = 1, \ldots, n$, with corresponding dose $dose_i$. Then, the probability that y_i takes on a value of 1 can be written as:

$$Prob(y_i = 1) = \frac{1}{1 + e^{-[\beta_0 + \beta_1 \times \log(\text{dose}_i)]}},$$

or equivalently:

$$\log\left[\frac{Prob(y_i = 1)}{1 - Prob(y_i = 1)}\right] = \beta_0 + \beta_1 \times \log(\text{dose}_i),$$

where β_0 is the intercept and β_1 is the slope. The logit model for binary data belongs to a class of models known as generalised linear models. The parameters are estimated using the method of maximum likelihood. See [14, 36] for more information on these methods.

As for continuous data, to obtain a meaningful estimate of the RP, a parallel model must be fit. This is similar to the straight-line relationship (Equation (6.3)), where the left-hand side of the equation is replaced with the logit of p:

$$\log\left[\frac{Prob(y_i = 1)}{1 - Prob(y_i = 1)}\right] = \beta_0 + [\beta_1 \times \log(\text{dose}_i)] + [\Delta_{\beta_0} \times x_i].$$

Here Δ_{β_0} is the intercept difference (test - standard) and x_i is binary indicator variable as previously defined. Figure 6.10 shows the logit fits to the example data set in Table B.7, and Table 6.8 summarises the parameter estimates for the parallel model.

Points to note

For a dose-response relationship to be fitted to the data, at least two doses with intermediate proportions (i.e., not at 0 or 1) of subjects with response 1 are required. Otherwise the slope parameter is undefined. The problem is similar to the 4PL case, discussed in Section 6.1.2. In this case it is advisable to review the choice of doses.

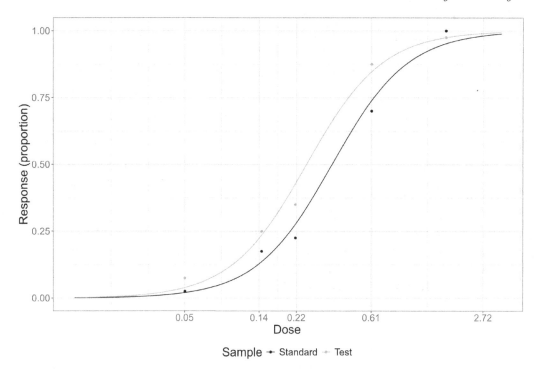

FIGURE 6.10
Plot of the parallel logit model for the data in Table B.7.

6.3 Combining relative potencies

If a bioassay method produces a measurement that does not have adequate precision for a particular procedure – for example batch release – then the method must be repeated and the results of the independent repeats combined. The result of the combination is known as the reportable value (RV), and it (rather than the individual assay results) is then compared to any applicable acceptance criteria such as the specification limits for a release assay.

We describe two approaches to the combination of measurements from repeated runs of a method; more information can be found in [59, 17]. A reportable value and its confidence interval are calculated in each case. The confidence interval depends on the assumption that the measurements are independent. If they share features such as dilutions of the reference standard, then the confidence interval may be too narrow. For both methods, we assume that we have n independent measurements of $\log(\mathrm{RP})$: r_i, each with a 95% confidence interval $(\mathrm{LCL}_i,\ \mathrm{UCL}_i)$, $i = 1, \ldots, n$.

6.3.1 Unweighted combination

The unweighted combination approach assumes all the measurements to be combined are equally important. It is based on the measurements alone and does not involve their individual confidence intervals.

The combined estimate of log(RP) is the mean value of the individual estimates:

$$\bar{r} = \frac{\sum_{i=1}^{n} r_i}{n}.$$

The standard deviation of the individual estimates is:

$$s = \sqrt{\frac{1}{n-1} \sum_{i=1}^{n} (r_i - \bar{r})^2}.$$

A $100(1-\alpha)\%$ confidence interval for log(RP) is then:

$$\bar{r} \pm t_{1-\frac{\alpha}{2}, n-1} \frac{s}{\sqrt{n}},$$

where $t_{1-\frac{\alpha}{2}, n-1}$ is the $1-\frac{\alpha}{2}$ percentile of the t distribution with $(n-1)$ degrees of freedom.

On the original scale, the combined estimate of RP is given by antilog $(\bar{r})$, which is equal to the geometric mean of the n measurements of RP. A $100(1-\alpha)\%$ confidence interval for RP is given by the back-transformed confidence interval for log(RP):

$$\frac{\text{antilog}\,(\bar{r})}{\text{antilog}\left(t_{1-\frac{\alpha}{2}, n-1} \frac{s}{\sqrt{n}}\right)} \;, \text{antilog}\,(\bar{r}) \times \text{antilog}\left(t_{1-\frac{\alpha}{2}, n-1} \frac{s}{\sqrt{n}}\right).$$

6.3.2 Weighted combination

The weighted combination approach assigns weights to the measurements according to the widths of their individual confidence intervals combined with the variance between the estimates. The wider the confidence interval, the smaller the weight for the measurement in the combination. There are two options when it comes to implementing a weighted approach: (i) to assume the presence of variability among the assay results due to differences in laboratory conditions; and (ii) to assume this variability is absent. In this section we only consider the former as it is the more conservative approach, it is often difficult to justify the absence of variability.

Like the unweighted combination, the calculations are conducted on the log scale. The weights, w_i, are based on the variance of r_i, which is the sum of the within-assay and between-assay variances. The within-assay variance for r_i is estimated by:

$$s_i^2 = \frac{(\text{UCL}_i - \text{LCL}_i)^2}{4 \times t_i^2},$$

where t_i is the t distribution quantile that was used to calculate the confidence limits. The average within-assay variance is therefore:

$$s_W^2 = \frac{1}{n} \sum_{i=1}^{n} s_i^2.$$

The between-assay variance can be calculated from the values of r_i as:

$$s_B^2 = \max\left(0, \frac{1}{n-1} \sum_{i=1}^{n} (r_i - \bar{r})^2 - s_W^2\right).$$

TABLE 6.9

Four estimates of RP for a given test sample, along with the combined RP estimates for both the unweighted and weighted methods.

	Relative potency	
	Estimate	95% CI
Individual session estimates		
1	0.930	(0.827, 1.047)
2	0.917	(0.736, 1.141)
3	1.076	(0.863, 1.341)
4	0.982	(0.841, 1.147)
Combined estimate		
Unweighted	0.974	(0.868, 1.093)
Weighted	0.960	(0.892, 1.035)

The weight for r_i is given by the reciprocal of its variance:

$$w_i = \frac{1}{(s_i{}^2 + s_B^2)}.$$

The combined estimate of log(RP) is the weighted mean value of the individual estimates:

$$\bar{r}_w = \frac{\sum_{i=1}^{n} w_i r_i}{\sum_{i=1}^{n} w_i}.$$

The standard error of the estimate is given by:

$$SE(\bar{r}_w) = \frac{1}{\sqrt{\sum_{i=1}^{n} w_i}}.$$

A $100(1 - \alpha)\%$ confidence interval for the reportable value on the log scale is then:

$$\bar{r}_w \pm t_{1-\frac{\alpha}{2}, n-1} \times SE(\bar{r}_w).$$

The value of $t_{1-\frac{\alpha}{2}, n-1}$ is often approximated by 2. On the original scale, the estimate of RV and its confidence interval are given by the back-transformed values.

Example: combining relative potencies

Table 6.9 shows an example set of four individual RP measurements collected from four separate bioassay sessions; Figure 6.11 shows a plot of the same data. The combined estimates for both the unweighted and weighted methods are also presented. The widths of the confidence intervals for the combined methods are narrower than the confidence intervals for the individual estimates reflecting the impact of replication on improving precision. The estimate for session 3 was somewhat higher than the estimates for the other three sessions. For the unweighted combination, this value has has pulled the final combined estimate towards itself. However, for the weighted method, since the confidence interval for the RP for

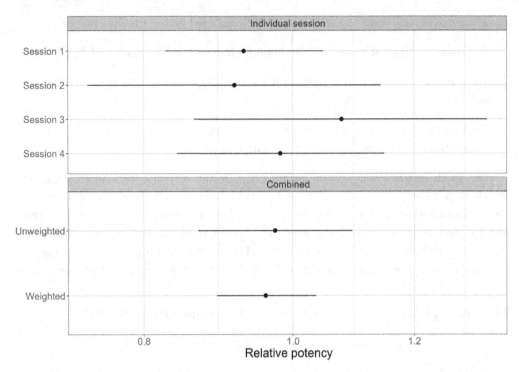

FIGURE 6.11

Four estimates of RP for a given test sample, along with the combined RP estimates for both the unweighted and weighted methods. The width of the confidence interval for the combined methods are narrower than the confidence intervals for the individual estimates reflecting the impact of replication on improving precision.

session 3 was also quite wide, its influence in the combined estimate is down-weighted and so contributes less to the final overall combined estimate. The width of confidence interval for the weighted method is also narrower than for the unweighted method.

6.4 Accounting for structure of replicates

When a bioassay method involves replication, of course the replicates will not produce the same response values. Some of the variability amongst the replicates may be explained by the fact that they occupy different parts of the plate, or even different plates (or cages or racks in the case of animal assays).

As a simple example, suppose a bioassay includes one replicate of the reference standard and one replicate of a test sample on a plate, and that four such plates make up the method. Also suppose that the dose-response relationship is a straight line. The method requires a single estimate of the relative potency.

There are several ways to calculate the RP combined across the four plates:

1. Calculate a value per plate and combine the four values (see Section 6.3).

2. Pool all the data together as though from a single plate (with four replicates) and calculate a single estimate.

3. Pool all the data together and account for the fact that there are four different plates, each estimating the same underlying relative potency.

For all three methods, we assume that y_{ijk} is the response for the ith sample ($i =$ standard, test) at the jth dose on the kth plate and x_{ijk} is an indicator variable that takes on a value of 0 if the sample corresponding to the ijk^{th} observation is the reference standard and 1 if the test sample. We also assume that ϵ_{ijk} is a random error term that is a normally distributed with mean zero and variance σ^2 and represents the variability of responses about the model. Then, the mathematical models for the data for the three methods are as follows.

Method 1, plates analysed separately and subsequently combined:

$$y_{ijk} = \beta_{0Sk} + [\Delta_{\beta_0 k} \times x_{ijk}] + [\beta_{1k} \times \log{(\text{dose}_{ijk})}] + \epsilon_{ijk},$$

where, for plate k:

β_{0Sk} is the intercept of the reference standard,

$\Delta_{\beta_0 k}$ is the difference between the intercepts, test - standard,

β_{1k} is the common slope for both samples.

The log relative potency for plate k is estimated as:

$$\log{\widehat{\text{RP}}_k} = \frac{\hat{\Delta}_{\beta_0 k}}{\hat{\beta}_{1k}}.$$

Method 2, all data analysed as if collected on single plate:

$$y_{ijk} = \beta_{0S} + [\Delta_{\beta_0} \times x_{ijk}] + [\beta_1 \times \log{(\text{dose}_{ijk})}] + \epsilon_{ijk},$$

where:

β_{0S} is the (mean) intercept of the reference standard,

Δ_{β_0} is the difference between the intercepts, test - standard,

β_1 is the common slope for both samples.

The log relative potency is estimated as:

$$\log{\widehat{\text{RP}}} = \frac{\hat{\Delta}_{\beta_0}}{\hat{\beta}_1}.$$

Note that this model assumes that there is no difference in the slopes and intercepts among the plates. However, if this is not the case and there are plate specific differences in the parameters, then this will violate the assumptions of the model and the confidence interval for the estimate will be wider than necessary.

Method 3, mixed model for combined data:

$$y_{ijk} = [\beta_{0S} + \delta_{\beta_0 k}] + [\Delta_{\beta_0} \times x_{ijk}] + [(\beta_1 + \delta_{\beta_1 k}) \times \log(\text{dose}_{ijk})] + \epsilon_{ijk},$$

where:

β_{0S} is the intercept of the reference standard,

$\delta_{\beta_0 k}$ is a normally distributed random term with mean zero and variance $\sigma_{\beta_0}^2$ which represents the variability in the intercepts for the standard across plates.

Δ_{β_0} is the difference between the intercepts, test - standard,

β_1 is the common (mean) slope for both samples,

$\delta_{\beta_1 k}$ is a normally distributed random term with mean zero and variance $\sigma_{\beta_1}^2$ which represents the variability in the slopes across plates.

The log relative potency is estimated as:

$$\log \widehat{\text{RP}} = \frac{\hat{\Delta}_{\beta_0}}{\hat{\beta}_1}.$$

This mixed model will result in the same estimate of relative potency as Method 2. However, since it accounts for the plate-to-plate to variability of the intercept of the reference standard and the variability of the slope, the confidence interval will tend to be narrower. We have not placed a random term on the intercept difference, Δ_{β_0}. This is because when the RP is fixed, a given difference in the intercept of the standard and a given slope imply a particular value of Δ_{β_0}.

In all cases, the 95% confidence interval for the estimate can be given by Fieller's Theorem. Note that, for Method 1, because the data for each plate are analysed separately, the estimate of the error variance will differ from plate to plate. However, for Methods 2 and 3, the error term has a variance which is not plate dependent; it can be thought of as an average of the variances across the plates.

For Method 1, as Fieller's Theorem is applied to each plate separately, the degrees of freedom for obtaining the CI of the RP are given in Section 6.1.1.1. For Method 2, the data are also analysed as per Section 6.1.1.1, but where the sample size, n, in the calculation of the degrees of freedom is taken as the total sample size summed across all plates. For Method 3, the degrees of freedom for the mixed model are a subject of ongoing debate and their estimate is not implemented in the lme4 [2] package we used to fit the data. For large sample sizes, the t-value in Fieller's Theorem could be approximated with a value of 2; however in the context of bioassay this may not be appropriate.

Examples: accounting for structure

We present two examples, where we compare the analysis of a bioassay using the three methods described above. In the first example, we assume that the slopes are constant across all plates, but that the intercepts may differ. In the second example, both the slope and the intercept are assumed to vary from plate to plate.

The differences between the three analysis methods depend on the extent of the differences between the plates. In Method 2, the differences between plates are not accounted for, so, if these effects are present, the estimate of potency will be less precise. If plate effects are suspected, then it is wise to use Method 1 or Method 3.

Example 1: constant slope across plates

In this first example, we assume that the results from 4 independent plates will be combined to form the final reportable value. The data for this example are found in Table B.8. Table 6.10 summarises the parameter estimates for the three methods including the estimate of the combined RP and its confidence interval.

Figure 6.12 plots the data showing the fitted straight-line relationship for Methods 2 and 3. For Method 3, we assumed that the slopes were constant across all plates, that is, that the variance of $\delta_{\beta_1 k}$ was 0. Consequently, we fitted a model where this term was ignored. The degrees of freedom for calculating the confidence interval for the RP were conservatively estimated to be $n - p$ (as per Equation (6.2)), where $n = 32$ is the total sample size across all plates, and the number of parameters $p = 6$.

From Table 6.10, the relative potency estimates are very similar for all methods except for Method 1, weighted method. The most precise results (narrowest confidence interval) was Method 1, weighted. The confidence intervals for Method 1, unweighted and Method 3, the mixed model, were similar. Since there were plate effects, the confidence interval for Method 2 was very wide.

Example 2: varying slope across plates

In this second example, we again assume that the results from 4 independent plates will be combined to form the final reportable value. The data for this example are found in Table B.9. Figure 6.13 plots these data along with the best fit lines to each individual plate and sample. Table 6.11 summarises the parameter estimates for the three methods including the estimate of the combined RP and its confidence interval.

Figure 6.12 plots the data showing the fitted straight-line relationship for Methods 2 and 3. For Method 3, we assumed that the slopes were different across all plates.

From Table 6.11, the relative potency estimates are very similar for all methods except for Method 1, unweighted method. The most precise results (narrowest confidence interval) was Method 3, mixed model. The confidence interval for Method 1, weighted was also reasonably precise. Again, since there were plate effects, the confidence interval for Method 2 was very wide.

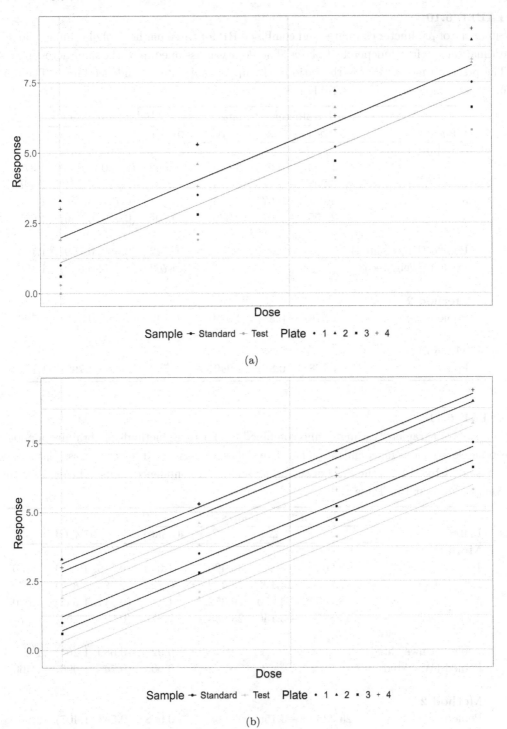

(a)

(b)

FIGURE 6.12

Plots of the straight-line relationship fit for Methods 2 in (a) and Method 3 in (b). It can be seen that between plates the intercepts differ, but that the slope is the same. The data for this example are presented in Table B.8.

TABLE 6.10

Summary of parameter estimates and combined RP for three methods of obtaining a pooled estimate across four independent plates. The slope was assumed to be the same across plates. The precision factor (PF) is the ratio of the upper and lower confidence limits. The data for this example are found in Table B.8.

Plate(s)	Parameter estimates			Relative potency		
	β_{0S}	Δ_{β_0}	β_1	Estimate	95% CI	PF
Method 1						
1	1.083	−0.925	3.095	0.742	(0.630 ,0.868)	1.379
2	3.208	−0.875	2.878	0.738	(0.611, 0.884)	1.448
3	0.727	−0.675	2.835	0.788	(0.727, 0.853)	1.173
4	2.895	−1.050	2.986	0.704	(0.536, 0.906)	1.691
Pooled (Unweighted)				0.742	(0.689, 0.799)	1.160
Pooled (Weighted)				0.769	(0.731, 0.809)	1.106
Method 2						
Pooled	1.978	−0.881	2.949	0.742	(0.553, 0.976)	1.766
Method 3						
Pooled	1.978	−0.881	2.949	0.742	(0.691, 0.795)	1.151

TABLE 6.11

Summary of parameter estimates and combined RP for three methods of obtaining a pooled estimate across four independent plates. The slopes were assumed to vary across plates. The precision factor (PF) is the ratio of the upper and lower confidence limits. The data for this example are found in Table B.9.

Plate(s)	Parameter estimates			Relative potency		
	β_{0S}	Δ_{β_0}	β_1	Estimate	95% CI	PF
Method 1						
1	20.905	1.400	33.420	1.043	(0.715, 1.523)	2.130
2	27.808	−14.125	43.422	0.722	(0.593, 0.877)	1.479
3	8.765	3.250	29.352	1.117	(0.811, 1.542)	1.900
4	43.825	−3.150	23.985	0.877	(0.593, 1.290)	2.174
Pooled (Unweighted)				0.927	(0.680, 1.264)	1.860
Pooled (Weighted)				0.905	(0.747, 1.096)	1.466
Method 2						
Pooled	25.326	−3.156	32.544	0.908	(0.583, 1.407)	2.412
Method 3						
Pooled	25.326	−3.156	32.544	0.908	(0.773, 1.051)	1.360

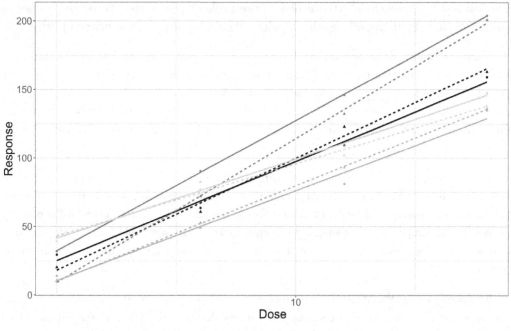

FIGURE 6.13

Plot of the dose-response data in Table B.9. The best fit lines to each individual plate and sample are shown. The slopes and intercepts clearly differ from plate to plate.

6.5 Outliers

In contrast with the discussion of outliers in Chapter 5, the identification and treatment of extreme observations or outliers for routine bioassay runs must be objective and fully specified.

There is always the possibility of outlying, or extreme, data values appearing in data sets by random chance alone. Such values should not be excluded from the data set: they represent the variability of the assay. Removing them is likely to introduce bias rather than remove it. Therefore, great care must be taken not to exclude data points unless there is evidence that they truly do not belong in the data set.

Ideally, all data points which do not belong in the data set because of human or instrument error would be identified and excluded on scientific grounds. In practice, this is not often possible. The next best thing is to flag extreme data points via one of the available statistical methods, then make a scientifically driven decision on the exclusion of each flagged data point. However, the scientific decision is usually still difficult or impossible.

The last resort is to exclude extreme observations solely because they have been flagged as outliers on statistical grounds. Such exclusions should be rare and the process justified.

There are many different statistical methods for flagging extreme data values. Most are based on the assumption that the data are normally distributed. This means these tests are not appropriate for binary data. We will discuss two such methods here: Grubbs' test and externally studentised residuals [25, 42]. In both cases, if the response is to be transformed for analysis, then this outlier analysis should also be based on the transformed response values.

Outliers can also be accommodated using techniques that are not as sensitive to their presence. These methods are known as robust statistics. An example of a robust statistic that provides a measure of central tendency is the median. The median is less influenced by outliers when compared to the mean. Analogous methods are available for the calculation of potencies but are beyond the scope of this book. For a more thorough discussion on robust statistics see [35].

The coefficient of variation (CV) is an often used, but inappropriate method, for flagging the presence of outliers in this context. We further describe the CV and the pitfalls of its use in bioassay in Chapter 7.

6.5.1 Grubbs' test

We first consider Grubbs' test [25] for the identification of outliers. This test assumes a data set $\{y_1, \ldots, y_n\}$ and has the following null (H_0) and alternative (H_A) hypotheses:

H_0: the data points all belong to the same distribution;

H_A: the data points do not all belong to the same distribution.

The Grubbs' test statistic is given by:

$$G = \frac{\max_{i=1,\ldots,n} |y_i - \bar{y}|}{s},$$

where $\bar{y}$ and s are the mean and standard deviation of the data set. The data point with the maximum value of $|y_i - \bar{y}|$ is flagged as extreme (an outlier) if G is greater than the critical value G_{critical}:

$$G > G_{\text{critical}} = \frac{n-1}{\sqrt{n}} \times \sqrt{\frac{t^2_{\frac{\alpha}{2n}, n-2}}{n-2+t^2_{\frac{\alpha}{2n}, n-2}}},$$

where α is the level of the test (usually 0.05 or 0.01), and $t_{\frac{\alpha}{2n}, n-2}$ is the $\left(1 - \frac{\alpha}{2n}\right)$ percentile of the t distribution with $n-2$ degrees of freedom.

If the most extreme point (i.e., the data point with the maximum value of $|y_i - \bar{y}|$) is flagged as extreme, then this point is excluded and the test repeated on the remaining points. This process can be repeated until there are only 2 points left.

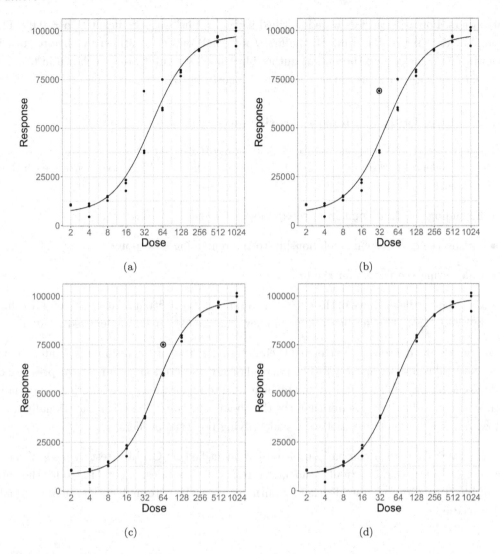

FIGURE 6.14

Iterative application of Grubbs' test to the data in Table B.10. The complete data set is shown in (a); the first outlier identified by Grubbs' test is shown in (b); the second outlier identified by Grubb's test in shown in (c); and the data set with the outliers removed is shown in (d).

Grubbs' test can be applied to the individual dose groups. Another approach is to fit the dose-response relationship to the data for the sample and calculate the residuals for the data points Equation (6.1). In the absence of true outliers, the residuals are assumed to come from the same normal distribution with mean value zero. Grubbs' test can be applied to the whole set of residuals at once. As above, the test can be repeated until there are no more extreme points.

Figure 6.14 shows an iterative application of Grubbs' test to the residuals from the 4PL fit to the data in Table B.10. This is an illustrative example that contains a high proportion of outliers. Potential outliers were identified if the Grubbs' was significant at $\alpha = 0.05$. The complete data set is shown in (a); the first (potential) outlier identified by Grubbs' test is shown in (b); the second (potential) outlier identified by Grubb's test in shown in (c); and the data set with both the outliers removed is shown in (d).

6.5.2   Externally studentised residuals

Another approach to flagging values as extreme (potential outliers) is to calculate the externally studentised residual for each point individually [42]. This involves, for each data point in turn:

- Excluding the data point from the data set;

- Fitting the dose-response relationship to the remaining data points;

- Calculating the RMSE for the fit;

- Calculating the distance of the excluded data point to the fitted relationship and dividing by the RMSE. This resulting value is known as the externally studentised residual.

Data points are identified as extreme if their externally studentised residual is greater than a pre-specified threshold, usually 3. The removal of any outlier(s) may unmask the presence of additional outlier(s). Therefore, it can be beneficial to repeat this process until no externally studentised residuals are greater than the threshold. This approach is computationally slow. However, it is the preferred method for identifying potential outliers.

Figure 6.15 shows a comparison of outliers identified by Grubbs' test at a significance level of 0.05 in (a) and externally studentised residuals at a threshold of 3 in (b) for the data in Table B.10. Grubbs' test identified 2 outliers, while the externally studentised residuals identified 5.

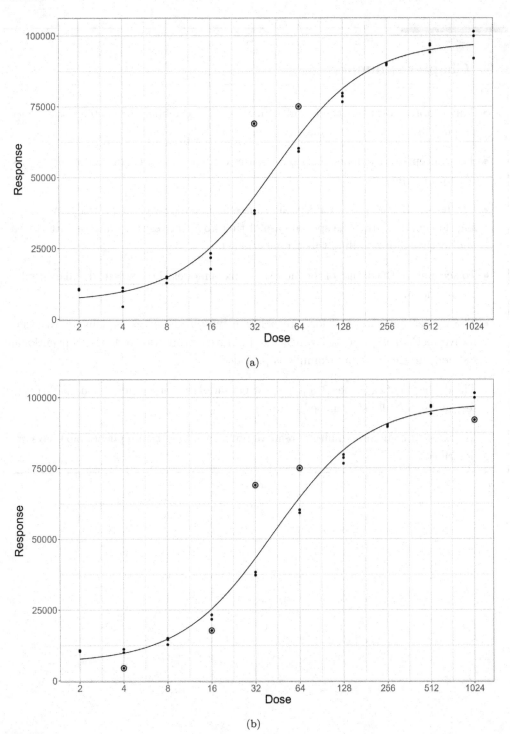

(a)

(b)

FIGURE 6.15

Comparison of outliers identified by Grubbs' test at a significance level of 0.05 in (a) and externally studentised residuals at a threshold of 3 in (b) for the data in Table B.10. Grubbs' test identified 2 outliers, while the externally studentised residuals identified 5.

6.6   Chapter summary

- This chapter covers methods for estimation of the model parameters, and hence potency, from the data; confidence intervals are also covered.

- The concepts of residuals and least squares estimates, and the use of iterative procedures, are introduced.

- The linear, 4PL and 5PL relationships are examined, as well as the logistic relationship for binary data. The special case of time-to-event response data is introduced and methods for handling this are mentioned.

- To account for the design of the bioassay, 'mixed models' are explained and examples are provided.

- The conventional method for interpolation-type bioassays is explained. An alternative method is presented which has improved properties over the conventional method. In each case, an example is provided.

- The detection of extreme data points, or potential outliers, is addressed via Grubbs' test and externally studentised residuals.

- Methods for combining multiple relative potency values from replicate bioassays are described.

7

Bioassay suitability criteria

Drawing statistical inferences from data should be accompanied by verification of the assumptions that have been made about the distribution of the data values and the relationships between variables. If the assumptions do not hold for the data set in hand, estimates of quantities such as relative potency can be biased and their variances can be incorrect. As a result, there is a risk that inappropriate conclusions about the materials being tested could be drawn.

We assume that the fundamental assumptions of homogeneity of variance, normality and independence have been met and that an appropriate dose-response model has be selected (see Chapter 5). Then, in the context of bioassay, once the statistical framework for estimating the potency has been identified, a set of criteria should be established so that, when a bioassay run is conducted, appropriate assessments can be made and hence the validity of the calculated result(s) can be established.

For all samples individually, whether the reference standard, quality control (QC) sample or a test sample: firstly, the presence of outlying observations may be determined; secondly, the fit of the dose-response relationship to the data should be verified. A third assessment should be made for similarity of the reference standard and the test (or QC) sample. This should be examined via the parallelism of the curves of the two samples. The presence of outliers, lack-of-fit and lack of parallelism can all lead to a decrease in the analytical performance of the assay. That is, they can result in bias of the estimate of potency and/or incorrect confidence intervals for model parameters (and potency).

These assessments on the data arising from a bioassay run fall into two logically distinct groups: those relating to the bioassay system and those relating to the test sample. We will refer to the criteria established for these groups as:

System suitability criteria: These criteria relate to the reference standard and the QC sample (if present). They provide assurance that the assay system behaved as expected. If any of the system suitability criteria are not met, then the results for all test samples are invalid and the whole bioassay needs to be repeated.

Sample suitability criteria: These criteria relate to the test sample. If all the system suitability criteria are met, but, for an individual test sample, any of the sample suitability criteria are not met, then the result for that test sample is invalid and the sample needs to be re-tested.

DOI: 10.1201/9781003449195-7

It is important to establish adequate system and sample suitability criteria, such that the bioassay can be relied upon to produce a result that is accurate. However, for any test, there is a chance of spurious failure, and the more tests that are implemented, the higher the chance of inappropriately declaring the result invalid. It is easy to impose too many criteria on the data, with the result that an unnecessarily high proportion of assays have to be repeated, sometimes multiple times, before a result is declared valid.

When developing a set of system and sample suitability criteria for a bioassay in routine use, it is important to keep the purpose in mind: to estimate the potency of the test sample accurately and precisely. The system and sample suitability criteria form part of the statistical analysis methodology for the bioassay. So long as the bioassay is demonstrated to be acceptably accurate and precise, according to the ATP, (see Chapter 2), then the system and sample suitability criteria which have been applied can be assumed to be adequately stringent.

7.1   Approaches to statistical tests: significance and equivalence

In this chapter we work within the frequentist statistical framework, where a 'null hypothesis', denoted by H_0, is tested. The probability of observing a data set at least as extreme as that observed is calculated, under the assumption that H_0 is true. This probability is known as the 'p-value'.

If the p-value is very small (less than a pre-stated value known as the significance level of the test and denoted by α, usually 0.05 or 0.01), then the conclusion is that the null hypothesis is inconsistent with the observed data and hence false. The complement of H_0 is known as the alternative hypothesis, H_A. If the conclusion is that H_0 is false then equivalently H_A is true. This can be thought of as a kind of 'proof by contradiction'. Conversely, if the p-value is greater than α, then the data observed are not inconsistent with H_0. This does not provide evidence that H_0 is true. Rather the data have merely failed to disprove H_0.

In this hypothesis testing framework, only the alternative hypothesis can be proved. The null hypothesis can be that two sets of data come from the same population – for example, they have the same value for the slope of the dose-response curve. The alternative is then that the two sets of data have different slopes. In terms of the model parameters, for the linear model, the hypotheses for a test of the difference between the slopes of the reference standard, β_{1S}, and test sample, β_{1T}, are:

$$H_0\colon \beta_{1S} = \beta_{1T}$$

$$H_A\colon \beta_{1S} \neq \beta_{1T}.$$

In this case the only conclusions that can be drawn are that either the two slopes are different or that there is not enough evidence to prove that they are different. It is not possible to conclude that the two slopes are the same. This is sometimes called a test of significance.

The converse of a significance test is a test of equivalence. In this case, the null hypothesis is that the slopes are different and the alternative is that they are the same (to within pre-stated bounds). For example, in the case of slopes for the linear model, this can be expressed in terms of the difference:

$$H_0: |\beta_{1S} - \beta_{1T}| \geq \delta$$

$$H_A: |\beta_{1S} - \beta_{1T}| < \delta,$$

where δ is the pre-stated boundary for equivalence. In the equivalence setting, the only conclusions that can be drawn then are that the two slopes are equivalent (or not meaningfully different) or that there is not enough evidence to prove that they are equivalent. It is not possible to conclude that the two slopes are different.

The α level equivalence test (i.e., where the probability of falsely claiming equivalence is α) can be conducted using the two-sided, $(1 - 2\alpha)$ confidence interval for $(\beta_{1S} - \beta_{1T})$. If the confidence interval lies within $(-\delta, \delta)$ then equivalence is demonstrated. Alternatively, two one-sided (significance) tests can be conducted. The hypotheses are:

$$H_{0(1)}: (\beta_{1S} - \beta_{1T}) \geq \delta; \quad H_{A(1)}: (\beta_{1S} - \beta_{1T}) < \delta,$$

$$H_{0(2)}: (\beta_{1S} - \beta_{1T}) \leq -\delta; \quad H_{A(2)}: (\beta_{1S} - \beta_{1T}) > -\delta.$$

If both null hypotheses are rejected then equivalence is demonstrated.

In the equivalence testing framework we usually test the two one-sided hypotheses at the 5% significance level. This corresponds to the two-sided 90% confidence interval falling inside the equivalence bounds $(-\delta, \delta)$.

An equivalence test relates to a pre-specified difference that is regarded as meaningful or important, whereas a significance test relies entirely on probability. Equivalence tests can be expressed for a range of metrics for the difference including: the arithmetic difference, the ratio, and the difference as a proportion of the mean value. Both approaches have their place and will be described in this chapter. The added complexity for an equivalence test is the setting of the equivalence bound, δ.

Figure 7.1 illustrates the two approaches in terms of a confidence interval for the difference between two slopes. For a significance test, a significant difference is concluded if the confidence interval does not contain zero. For an equivalence test, the results are equivalent if the confidence interval lies within the equivalence bounds.

7.2 Setting equivalence bounds

There are a number of methods that can be used to set the equivalence bounds. Ideally, when applied, an equivalence test should be able to discriminate between assays that are

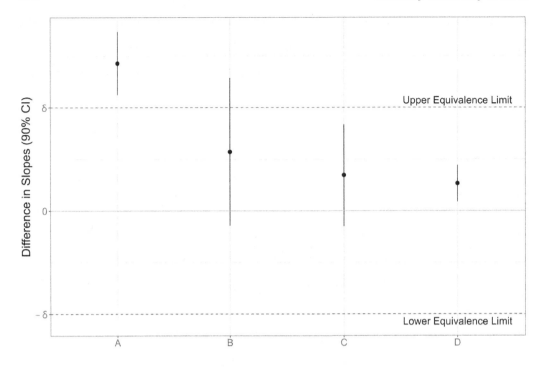

FIGURE 7.1

For a significance test, a significant difference is concluded if the confidence interval does not contain zero. For an equivalence test, the results are equivalent if the confidence interval lies within the equivalence bounds. The conclusions for the intervals are as follows: A: significantly different and not equivalent; B: not significantly different and not equivalent; C: not significantly different and equivalent; and D: significantly different and equivalent.

performing as expected (and therefore are likely to produce 'good' data) and assays that are not. For example, for an equivalence test assessing parallelism, the bounds should be chosen such that samples that are truly parallel are likely to pass, and samples that are not parallel are likely to fail.

If the equivalence bounds are set too narrow, the assay is likely to suffer from a high failure rate. If the equivalence bounds are set too wide, then it is likely to result in increased variability for the reported result, and/or the possibility of reporting an invalid RP. Therefore, the choice of data used to set the equivalence bounds is critical. It is important that any sources of variability that may occur in the routine use of the assay are reflected, but any additional sources of variability are not. Additional sources of variability may include changes to the assay as it undergoes development.

We assume that the equivalence bounds will be set based on a set of n results where, for each result, we have a confidence interval with lower and upper confidence limits denoted as (LCL_i, UCL_i), $i = 1, \ldots, n$. We assume that the confidence intervals for the chosen metric are directly given by statistical software. In general, the calculation of these confidence intervals is complex and cannot be easily obtained from the confidence intervals for the

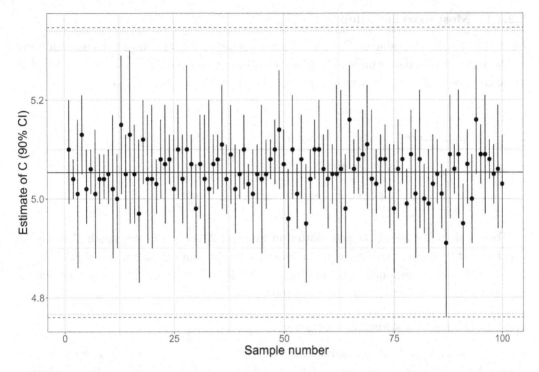

FIGURE 7.2

Equivalence bounds for the C parameter of a 4PL fit to the reference standard. The bounds have been set using (a) the most extreme observed value (dark grey dashed lines) and (b) the mean + 3 standard deviations (light grey dotted line). For these data the bounds for the two methods are nearly identical. These data can be found in Table B.11.

individual elements of the metric. We will look at two methods for obtaining equivalence bounds in detail: (i) bounds based on the most extreme observed value; and (ii) bounds based on the mean + 3 standard deviations.

How the equivalence bounds are set will depend on the parameter. For example, in setting equivalence bounds for the C parameter of the 4PL curve fit to the reference standard, one possibility is to set bounds symmetrically about the true value of C. Typically the true value of C will be unknown but can be estimated from the data.

Figure 7.2 shows a plot where equivalence bounds have been set symmetrically about the mean value of C across the 100 estimates. The bounds have been set using (a) the most extreme observed value and (b) the mean + 3 standard deviations. In this case the bounds for the two methods are nearly identical. The data for this plot can be found in Table B.11.

7.2.1 Individual parameter estimate

For an individual parameter, one option is to enforce symmetry around the estimate of its true value. We will illustrate the concept in terms of the C parameter for a 4PL fit to the reference standard. However, this could be any other parameter of interest and the choices available will depend on the model being fit.

7.2.1.1   Most extreme value

Let $\bar{C}$ be the average value of C across the n measurements. The most extreme value will be the furthest confidence limit from $\bar{C}$ across all n measurements. Let d_i be the maximal distance between $\bar{C}$ and the confidence limits for the ith measurement, i.e.:

$$d_i = \max \left\{ \left| LCL_i - \bar{C} \right|, \left| UCL_i - \bar{C} \right| \right\}.$$

Then, the equivalence bounds can be set as:

$$\text{Lower equivalence bound} = \bar{C} - \max_{i=1,\ldots,n} \{d_i\},$$

$$\text{Upper equivalence bound} = \bar{C} + \max_{i=1,\ldots,n} \{d_i\}.$$

For the data in Table B.11 and plotted in Figure 7.2, the parameter mean $\bar{C} = 5.0533$. The furthest value from this point is the lower confidence limit of sample number 87 which is 4.7600, with a corresponding distance of $d_{87} = 0.2933$. Therefore, the equivalence bounds are calculated to be $5.0533 \pm 0.2933 = (4.7600, 5.3466)$.

7.2.1.2   Mean + 3 standard deviations

The equivalence bounds for an individual parameter can be also be set using the mean and standard deviation of the maximal distance from the overall estimate of the parameter's true value, $\bar{C}$ in this example. Let $\bar{d}$ and s_d be the mean and standard deviation of the maximum distances, d_i, $i = 1, \ldots, n$. Then, the equivalence bounds can be set as:

$$\text{Lower equivalence bound} = \bar{C} - \left(\bar{d} + 3 \times s_d \right),$$

$$\text{Upper equivalence bound} = \bar{C} + \left(\bar{d} + 3 \times s_d \right).$$

For the data in Table B.11 and plotted in Figure 7.2, the parameter mean $\bar{C} = 5.0533$, and the mean and standard deviation of the maximal distances are $\bar{d} = 0.1290$ and $s_d = 0.0525$. Therefore, the equivalence bounds are calculated to be $5.0533 \pm (0.1290 + 3 \times 0.0525) = (4.7669, 5.3340)$.

7.2.2   Difference of two parameter estimates

When forming equivalence bounds for the difference of two parameters, a difference of 0 indicates the parameters are the same. For example, when the lower asymptotes of the reference standard, A_S, and test sample, A_T, are the same, then the corresponding difference $A_S - A_T = 0$. Therefore, the bounds for an equivalence test for a difference between two parameters should be symmetric about a value of 0.

7.2.2.1   Most extreme value

The most extreme value for a difference between two parameters will occur at the maximum absolute value across all observed confidence limits. Let m_i be the maximum absolute value of the confidence limits for the ith measurement:

$$m_i = \max \left\{ \left| LCL_i \right|, \left| UCL_i \right| \right\}.$$

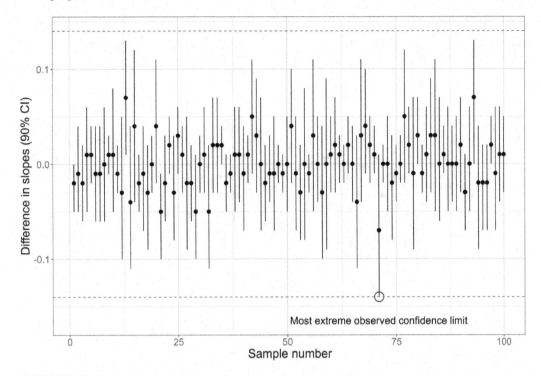

FIGURE 7.3

Differences in slope and 90% confidence interval for 100 bioassays. Equivalence bounds can be set based on the most extreme value. For these data, the most extreme limit corresponds to the lower confidence limit of sample number 71 with a value of -0.14. To ensure symmetry about a value of 0, the equivalence bounds are set at ± 0.14. The data for this plot can be found in Table B.12.

Then, the equivalence bounds can be set as:

$$\text{Lower equivalence bound} = -\max_{i=1,\ldots,n}\{m_i\},$$
$$\text{Upper equivalence bound} = +\max_{i=1,\ldots,n}\{m_i\}.$$

Figure 7.3 demonstrates this for the samples known to be parallel in the data sat provided in Table B.12. For these data the most extreme limit corresponds to a lower confidence limit with a value of -0.14. To ensure symmetry about a value of 0, the equivalence bounds are set at ± 0.14.

7.2.2.2 Mean + 3 standard deviations

For a difference between two parameters, equivalence bounds symmetric about 0 can be obtained as follows. Let $\bar{m}$ and s_m be the mean and standard deviation of the n observed maximum absolute confidence limits, m_i, $i = 1, \ldots, n$. Then, the upper equivalence bound can be set at the mean + 3 standard deviations and the lower equivalence bound as negative

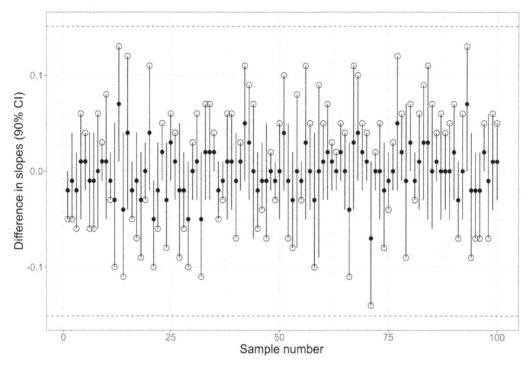

FIGURE 7.4

Differences in slope and 90% confidence interval (with limit further from 0 identified) for 100 bioassays. When setting equivalence bounds for a difference of two parameters, the bounds can be set using the mean and standard deviation of the confidence limit furthest from 0. The data for this plot can be found in Table B.12 of Appendix B.

upper bound.

$$\text{Lower equivalence bound} = -\left(\bar{m} + 3 \times s_m\right),$$
$$\text{Upper equivalence bound} = \left(\bar{m} + 3 \times s_m\right).$$

Figure 7.4 shows a plot with the limit furthest from 0 highlighted for the samples known to be parallel in the data set provided in Table B.12. For these data, $\bar{m} = 0.0667$ and $s_m = 0.0281$ so the equivalence bounds can be calculated to be $\pm(0.0667 + 3 \times 0.0281) = (-0.1509, 0.1509)$. These bounds are a little bit wider than those calculated using the most extreme value.

7.2.3   Ratio of two parameter estimates

When forming equivalence bounds for the ratio of two parameters, a ratio of 1 indicates that the parameters are the same. For example, if the slopes of the reference standard, β_S, and test sample, β_T, are the same, then their ratio $\frac{\beta_S}{\beta_T} = 1$. Therefore, the bounds for an equivalence test for a ratio of two parameters should have multiplicative symmetry about a value of 1. This can be achieved by first log transforming the confidence limits for the ratio, then using analogous calculations to those for a difference and back-transforming.

7.2.3.1 Most extreme value

The most extreme value will be the back-transformed maximum absolute value across the observed confidence limits on the log-scale. Let $m_{i,\log}$ be the maximum absolute value of the log-transformed confidence limits for the ith measurement:

$$m_{i,\log} = \max\left\{|\log LCL_i|, |\log UCL_i|\right\}.$$

Then, the equivalence bounds can be set as:

$$\text{Lower equivalence bound} = \text{antilog}\left[-\max_{i=1,\ldots,n}\{m_{i,\log}\}\right],$$

$$\text{Upper equivalence bound} = \text{antilog}\left[\max_{i=1,\ldots,n}\{m_{i,\log}\}\right].$$

7.2.3.2 Mean + 3 standard deviations

For a ratio of two parameters, equivalence bounds with multiplicative symmetry about 1 can be obtained as follows. Let $\bar{m}_{\log}$ and $s_{m,\log}$ be the mean and standard deviation of the n observed maximum absolute confidence limits on the log-scale, $m_{i,\log}$, $i = 1, \ldots, n$. Then, the upper equivalence bound can be set at the antilog of the mean + 3 standard deviations and the lower equivalence bound as antilog negative upper bound.

$$\text{Lower equivalence bound} = \text{antilog}\left[-\left(\bar{m}_{\log} + 3 \times s_{m,\log}\right)\right],$$

$$\text{Upper equivalence bound} = \text{antilog}\left[+\left(\bar{m}_{\log} + 3 \times s_{m,\log}\right)\right].$$

7.2.4 Other approaches

Other approaches to setting the equivalence bounds are available. These include tolerance intervals and percentiles. For more details on percentiles see Appendix A and Chapter 2 for more information on tolerance intervals.

7.2.5 Assessment of equivalence bounds

Before putting equivalence bounds into practice, it is important to plot these bounds against the data used to derive them, as well as any further assay data available. These additional data should ideally include samples that are expected to fail, for example, forced degradation samples. Figure 7.5 shows the data from Table B.12, but with the additional ten samples known to be non-parallel. Ideally, the bounds would be able to discriminate between the truly parallel and non-parallel samples. However, as is the case here, a trade-off will often need to be made as the bounds cannot be set such that only parallel samples pass and non-parallel samples fail.

7.3 Tests for individual samples

Two issues are relevant to individual samples: extreme observations and lack-of-fit of the dose-response relationship to the data.

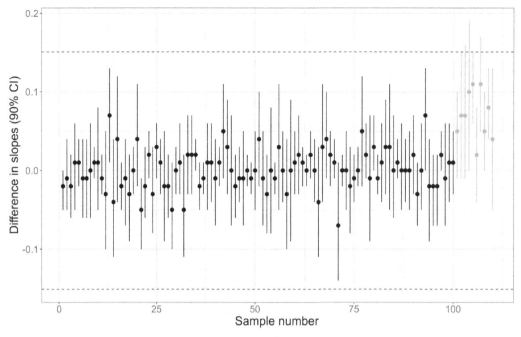

FIGURE 7.5

Plot of confidence intervals for difference in slopes including data for samples known to be truly parallel and samples known to be non-parallel. The equivalence bounds drawn as horizontal dotted lines are not able to (perfectly) discriminate between the two and a trade-off needs to be made. The data for this plot can be found in Table B.12 of Appendix B.

7.3.1   Extreme observations / outliers

The presence of extreme observations need not invalidate the assay. However, the method for handling these should be established for routine analysis. Chapter 6 provides details on methods for identifying extreme data values.

If many data values are excluded, the impact on the precision of the individual estimate may be considerable: this can be assessed via the confidence interval for the potency estimate (see Section 7.5).

A widely used suitability criterion is the coefficient of variation, CV. The CV for a variable is defined as the standard deviation, SD, divided by the mean:

$$CV = \frac{SD}{mean}.$$

If a log transformation has been applied this may no longer hold. A widely used, but, in the context of bioassay, inappropriate criterion imposed on treatment groups, sometimes with a view to finding extreme observations, is an upper bound on the CV. Bearing in mind that the SD must be constant across dose groups, coupled with the fact that when there is a dose-response relationship, the mean value is not constant across dose groups, the CV should increase as the mean value decreases. Therefore, imposing a single bound on the CV across all dose groups is likely to lead to bias because data will predominantly be dropped at low response levels.

7.3.2 Fit of the dose-response relationship

The dose-response relationship that is fitted to the data is the basis for the potency estimate. The assumption is that the fitted relationship is appropriate for the data, and this assumption should ideally be checked.

Model fit is particularly important in the case of a linear dose-response relationship, where curvature can also cause lack of parallelism. Where lack-of-fit is found, measures can be specified to find a subset of doses which result in a linear dose-response relationship.

While commonly used, the coefficient of determination, R^2, is not an ideal method for assessing the fit of the model in this context, as it assesses a combination of model fit and data variability.

There are several approaches to checking the fit of the dose-response relationship. One is based on an examination of the residuals and takes the significance approach, using an F-test. This test is only applicable for continuous response data and is very sensitive when data are precise, i.e., the pure error is low. Another is a model nesting approach whereby a parameter is added to the relationship being checked, and the improvement in the quality of the fit to the data is assessed. The latter approach can be framed either as a significance test or an equivalence test and can be applied to both continuous and binary response data.

7.3.2.1 Residuals-based test: continuous response only

Let $i = 1, \ldots, d$ index the dose groups of the data set and $j = 1, \ldots, n_i$ index the n_i independent replicates in dose group i. Then, the total number of observations can be found by summing over the number of replicates in each dose group $n = \sum_{i=1}^{d} n_i$. Let the jth response at dose i be denoted by y_{ij}. As described in Chapter 5, we assume y_{ij} is normally distributed and can be expressed as:

$$y_{ij} = f(\text{dose}_i) + \epsilon_{ij},$$

where f is the dose-response relationship (straight line, 4PL, 5PL) that when evaluated at dose i, $f(dose_i)$, gives the mean response, μ_i, and ϵ_{ij} is a normally distributed error term with mean 0 and variance σ^2. Let p be the number of parameters required to characterise the dose-response relationship, where $p = 2$ for a straight line relationship, and 4 and 5 for the 4PL and 5PL relationships, respectively. The variance of y_{ij}, σ^2, is assumed to be constant across all doses and represents the variability between an observed point and the line.

Let $\hat{y}_i$ be the fitted response for dose group i, where the subscript j may be omitted as all replicates within the same dose group will share a fitted response. The total variability of the individual observations about the fitted line can be characterised by a quantity known

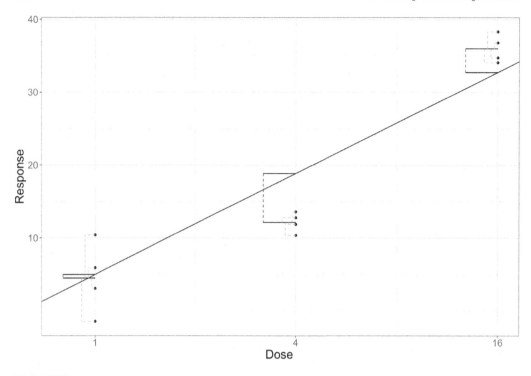

FIGURE 7.6

Straight-line fit to three dose groups showing the lack-of-fit error (black dashed lines) and pure error (grey dashed lines).

as the sum of squared errors, SSE (also known as the residual sum of squares previously defined). The SSE is calculated as the sum of the squared residuals:

$$\text{SSE} = \sum_{i=1}^{d} \sum_{j=1}^{n_i} (y_{ij} - \hat{y}_i)^2 ,$$

and will have degrees of freedom $n - p$. The mean square error, or MSE, corresponds to the SSE divided by its degrees of freedom.

The total variability about the fitted dose-response relationship can be broken down into two components: (i) the lack-of-fit error; and (ii) the pure error. The lack-of-fit error represents the variability of the dose group means about their fitted values. The pure error represents the variability of the response values of the individual dose replicates about their dose group mean. The lack-of-fit error could (at least in theory) be explained by fitting a more complicated dose-response relationship, while the pure error will always remain unexplained. Figure 7.6 provides a visualisation of the pure error and lack-of-fit error in the linear case – we explain these further below.

Let $\bar{y}_i$ be the average response value across the n_i replicates for the ith dose group. For a well-fitting dose-response relationship, the distance between this average and the fitted response (i.e., the line/curve), $\hat{y}_i$, should be small. The black dashed lines in Figure 7.6 represent the lack-of-fit of the relationship. The sum of squares due to lack-of-fit, SSLoF, can be calculated as:

$$\text{SSLoF} = \sum_{i=1}^{d} n_i \times (\bar{y}_{i\cdot} - \hat{y}_i)^2,$$

and will have degrees of freedom $d - p$. The mean square due to lack-of-fit, MSLoF, can be found as SSLoF divided by its degrees of freedom.

The pure error represents the variability of the dose replicates about their mean $\bar{y}_i$. This error is due only to random chance and is not related to the fit of the dose-response relationship. The sum of squares due to pure error, SSPE, can be calculated as:

$$\text{SSPE} = \sum_{i=1}^{d} \sum_{j=1}^{n_i} (y_{ij} - \bar{y}_{i\cdot})^2,$$

and will have degrees of freedom $n - d$. The mean square pure error, MSPE, is SSPE divided by its degrees of freedom. The grey dashed lines in Figure 7.6 represent the pure error.

It can be shown that SSE is the sum of these two components:

$$\text{SSE} = \text{SSLoF} + \text{SSPE}.$$

We can also show that the MSLoF, has expected value

$$\mathbb{E}(\text{MSLoF}) = \sigma^2 + \frac{\sum_{i=1}^{d} n_i \left[\mu_i^* - f(\text{dose}_i) \right]^2}{d - p}, \tag{7.1}$$

where μ_i^* is the mean response for the ith dose group and may not be given by the assumed dose-response relationship. For a dose-response model that fits the data perfectly, $[\mu_i^* - f(\text{dose}_i)] = 0$ and the right-hand term of Equation (7.1) will be 0. A hypothesis test to determine if the right-hand term is significantly greater than 0 can be conducted by exploiting the fact that the expected value of the MSPE is σ^2. The test statistic for this lack-of-fit F-test is:

$$\begin{aligned} F &= \frac{\sum_{i=1}^{d} n_i (\bar{y}_{i\cdot} - \hat{y}_i)^2 / (d - p)}{\sum_{i=1}^{d} \sum_{j=1}^{n_i} (y_{ij} - \bar{y}_{i\cdot})^2 / (n - d)} \\ &= \frac{\text{MSLoF}}{\text{MSPE}}. \end{aligned}$$

When the fitted dose-response relationship is satisfactory for the observed data, then the F-statistic will be close to a value of 1. If the fitted dose-response relationship is not satisfactory, then the F-statistic will be (much) larger than 1.

The lack-of-fit is deemed significant at level α if F is greater than the critical value $F_{critical}$:

$$F > F_{critical} = F_{1-\alpha, d-p, n-d},$$

where $F_{1-\alpha, d-p, n-d}$ is the $(1 - \alpha)$ percentile of the F-distribution with $(d - p)$ and $(n - d)$ degrees of freedom. This is equivalent to the p-value being less than α.

Example: straight-line relationship

The model fit is particularly important in the straight line case. Presence of curvature may indicate a departure from the 'linear part' of the dose-response curve and this can lead to bias in the potency estimate (see Chapter 4). When a curve is detected, it may be possible to find a contiguous subset of doses for which the relationship is linear, removing the need to re-run the assay.

Figure 7.7(a) shows an example of bioassay data, where a straight-line relationship is being used for the data analysis. These data can be found in Table B.13. Here there is clear curvature and the p-value for the lack-of-fit F-test is < 0.001, indicating significance at any reasonable level. However, when only the last four doses are chosen, as shown in Figure 7.7(b), a linear model fits the data well and there is no evidence of lack-of-fit. In this case, the p-value for the lack-of-fit F-test is 0.1921.

Example: S-shaped relationships

Figure 7.8(a) shows the 4PL model fit to the data found in Table B.14. There is asymmetry in the data and so this relationship does not fit the data well; the p-value for the lack-of-fit F-test is < 0.001. For this fit, there are some dose groups that sit entirely above or entirely below the fitted relationship. Figure 7.8(b) shows the same data set but with a 5PL model fitted. This model offers a much better fit to the data and there is no evidence of lack-of-fit with p-value for the F-test of 0.5925. There are no dose groups for the 5PL model that sit entirely above or entirely below the curve.

7.3.2.2   Adding a parameter to the relationship: continuous and binary data

Another approach to checking the fit of the dose-response relationship is to examine the improvement in the fit from adding a parameter to the dose-response relationship. This approach can be used for both continuous and binary data.

Straight line vs quadratic curve

The fit of a straight line can be assessed by comparing it with a curve. A simple curve is given by a quadratic dose-response relationship:

$$\mu_i = \beta_0 + \beta_1 \times \log\left(\text{dose}_i\right) + \beta_2 \times \left[\log(\text{dose}_i)\right]^2.$$

Here we have an extra parameter, β_2, which expresses a curve. Note that this relationship can only be concave upwards or downwards. It cannot express curves in both directions.

When $\beta_2 = 0$, the relationship is a straight line. Therefore, a test of the fit of the straight line relationship can be expressed as a test of the null hypothesis:

$$H_0: \beta_2 = 0,$$

versus the alternative hypothesis:

$$H_A: \beta_2 \neq 0.$$

This significance test can be conducted as a t-test (or the equivalent F-test), or as a likelihood ratio test [12].

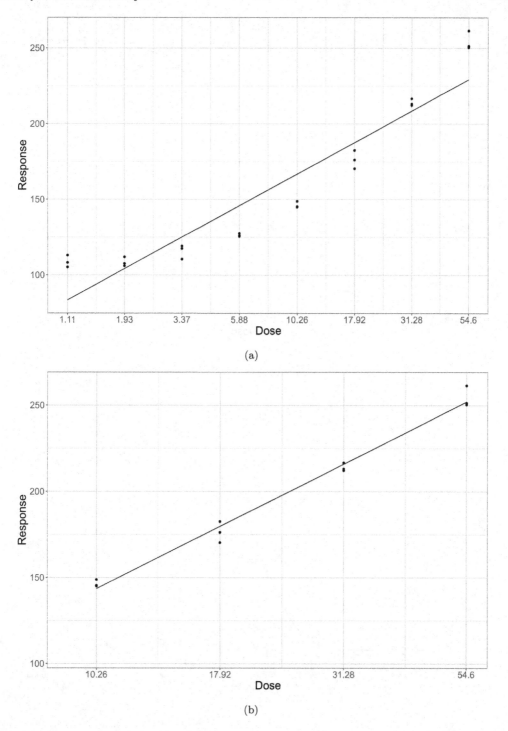

(a)

(b)

FIGURE 7.7

Lack-of-fit in a linear model with curvature in (a); when only the last four doses are used, then a linear model fits the data well and there is no evidence of lack-of-fit in(b). The data for this plot can be found in Table B.13.

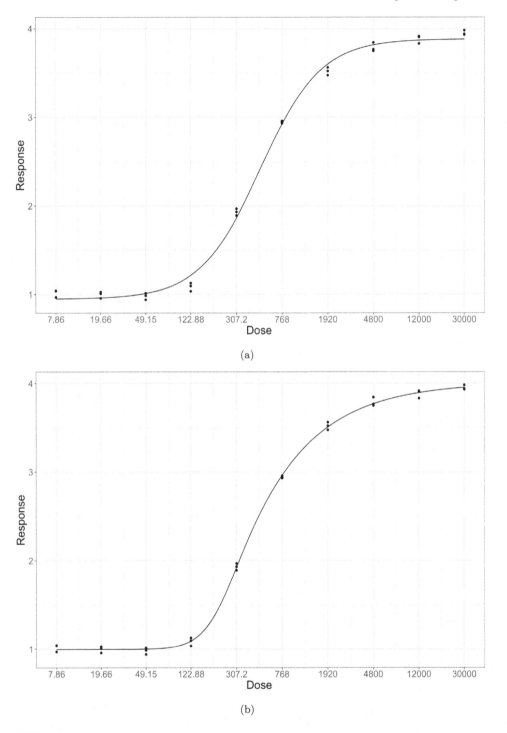

(a)

(b)

FIGURE 7.8

A 4PL fit to data that have asymmetry in (a); a 5PL fit to the same data provides a better fit in (b). For the 4PL fit, there are some dose groups that sit entirely above or entirely below the fitted relationship. The data for this plot are found in Table B.14.

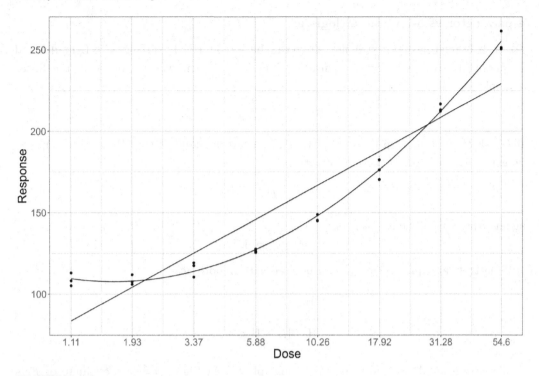

FIGURE 7.9
Quadratic fit to the data shown in Table B.13. The fit to the data has been greatly improved by adding the quadratic term.

Alternatively an equivalence test can be used:

$$H_0: |\beta_2| \geq \delta$$

$$H_A: |\beta_2| < \delta.$$

The value of δ should be chosen based on the effect of a departure from a straight line on the potency estimate. However, this is difficult to determine, and a data driven value is usually chosen. See Section 7.2.1 for examples.

The data in Table B.13 (and plotted in Figures 7.7 and 7.9) were examined for curvature by fitting a quadratic relationship. The estimate of the quadratic term $\beta_2 = 11.99$ with a 99% confidence interval of (10.46, 13.43). Using the significance approach, the null hypothesis $H_0: \beta_2 = 0$ is rejected at the 1% level (since the 99% confidence interval does not include 0). Therefore, we conclude that the data are not consistent with a straight line model. Figure 7.9 shows a plot of the quadratic fit, where we can see that the fit to the data has been greatly improved.

Using the equivalence approach, with, for example, $\delta = 5$, the null hypothesis is $H_0: |\beta_2| \geq 5$. The estimate of β_2 was 11.99 with a two-sided 90% confidence interval (11.06, 12.92). Since the 90% confidence interval for β_2 is entirely above δ we have failed to demonstrate that data are consistent with a straight line model.

4-parameter vs 5-parameter logistic

For a 4PL relationship, extra flexibility can be introduced by fitting a 5PL model and examining the value of the 5th parameter, E, in a similar way as for β_2 in the linear example. For dose i:

$$\mu_i = D + \frac{A - D}{\left[1 + e^{B(\log(dose_i) - C)}\right]^E}.$$

If $E = 1$, the relationship becomes 4PL. Thus, a test of the fit of the 4PL relationship can be expressed as a test of the null hypothesis:

$$\text{H}_0\text{: } E = 1,$$

against the alternative hypothesis:

$$\text{H}_A\text{: } E \neq 1.$$

Alternatively an equivalence test can be used with hypotheses:

$$\text{H}_0\text{: } |E - 1| \geq \delta$$

$$\text{H}_A\text{: } |E - 1| < \delta.$$

For the example data in Table B.14 and plotted in Figure 7.8, the estimate of E is 0.2178 with a 99% confidence interval (0.1365, 0.3477). Since the confidence interval does not contain 1, the data are demonstrated to be inconsistent with a 4PL model at the 1% level. That is, the 4PL model does not fit the data well.

Using the equivalence approach, with, for example, $\delta = 0.5$, the null hypothesis is $\text{H}_0\text{: } |E - 1| \geq 0.5$. Therefore, we will reject the null hypothesis (and accept the 4PL fit) if the 90% confidence interval for E lies entirely within the interval (0.5, 1.5). The 90% confidence interval for the E parameter of these data is (0.1636, 0.2901) and falls entirely below this interval. Therefore, from an equivalence perspective we have failed to demonstrate that data are consistent with a 4PL model.

7.3.2.3   Fit of the dose-response relationship: binary data

Where each individual unit results in a binary response the response is not normally distributed. Therefore, the residuals-based methods described above do not apply.

The linearity of the logit relationship with respect to $\log(dose_i)$ can be checked by adding a quadratic term as demonstrated in Section 7.3.2.2 above. For binary data, this corresponds to:

$$\log\left(\frac{p_i}{1 - p_i}\right) = \beta_0 + \beta_1 \times \log(dose_i) + \beta_2 \times [\log(dose_i)]^2$$

This analysis is relatively complex and requires generalised linear models, and we do not include further details.

TABLE 7.1

Reference standard parameters for the continuous dose-response relationships (straight-line, 4PL, 5PL) and binary dose-response relationship (straight line) we have discussed.

	Continuous data			Binary data
Parameter	Straight line	4PL	5PL	Straight line
Location	β_0	C_S	C_S	β_0
Slope parameter	β_1	B_S	B_S	β_1
Left asymptote		A_S	A_S	
Right asymptote		D_S	D_S	
Range (asymptote difference)		$(D_S - A_S)$	$(D_S - A_S)$	
Asymmetry			E_S	

7.3.3 Values of fitted reference standard parameters

If the reference standard has not changed, it should always produce roughly the same dose-response curve in the same bioassay system. Therefore, system suitability tests based on the reference standard dose-response curve parameters – slope, asymptotes, intercept, log EC_{50} etc – can be used to check that the bioassay is behaving as expected. These tests can be framed as either a significance or equivalence test. If these are not within expected ranges, the assay system may not be functioning correctly. Table 7.1 lists the parameters for the continuous and binary models we have discussed.

When setting equivalence bounds for the parameters, they should be set based on accumulated data from repeated bioassays. As more data accumulate, the bounds can be updated but care should be taken to ensure that the reference standard is stable (see Chapter 9). Bounds can be applied either to the estimate of the parameter or to a confidence interval for the parameter (usually 90% two-sided). The latter, equivalence, approach captures the precision of the individual run as well as the value of the parameter. An example of the equivalence bounds for the C parameter of the 4PL relationship is shown in Figure 7.2.

7.4 Tests for pairs of samples: parallelism

The underlying assumption of parallelism is perhaps the most important to assess in the conduct of a bioassay. The concept of a potency measure relative to a reference standard is not meaningful without it.

Where the dose-response relationship is a straight line (for either continuous or binary data), parallelism means that the two samples have equal slopes: $\beta_{1S} = \beta_{1T}$. Where the dose-response relationship is 4PL, parallelism means that the two samples have equal slope parameters and equal asymptotes: $B_S = B_T, A_S = A_T, D_S = D_T$. For the 5PL relationship, the two samples additionally have the same asymmetry parameter, $E_S = E_T$.

We consider two approaches for investigating parallelism of a model: an F-test, based on residuals, and an equivalence test approach. As for the lack-of-fit F-test, the F-test for parallelism is very sensitive when the data are precise, i.e., the pure error is low. Small departures from parallelism can produce a very low p-value but may not have an important effect on the potency estimate. However, the F-test can be useful in the early stages of development but, for routine use, equivalence testing is preferable. Equivalence testing requires a body of data for the test parameter in order to set the bounds.

7.4.1   Significance tests for parallelism: continuous data

Let $i = 1, \ldots, n$ index the observations, where n is the total number of observations across the reference standard and test samples. Then let y_i be the observed response for the ith observation with corresponding dose given by dose_i. Finally let x_i be a binary indicator variable that takes on a value of 0 if the sample corresponding to the ith observation is the reference standard and 1 if the test sample. We assume that the responses for the reference standard and test sample are normally distributed, with variance σ^2 across both samples and all dose groups. Therefore, the error term, ϵ_i, is normally distributed with mean 0 and variance σ^2.

The separate (unconstrained) straight-line relationship is defined as:

$$y_i = \beta_{0S} + [\beta_{1S} \times \log(\text{dose}_i)] + [\Delta_{\beta_0} \times x_i] + [\Delta_{\beta_1} \times \log(\text{dose}_i) \times x_i] + \epsilon_i, \tag{7.2}$$

while the parallel (constrained) straight line relationship is defined as:

$$y_i = \beta_{0S} + [\beta_1 \times \log(\text{dose}_i)] + [\Delta_{\beta_0} \times x_i] + \epsilon_i. \tag{7.3}$$

For the 4PL relationship, the unconstrained model is given by:

$$y_i = \begin{cases} D_S + \dfrac{A_S - D_S}{1 + e^{B_S[\log(\text{dose}_i) - C_S]}} + \epsilon_i, & x_i = 0 \text{ (i.e., reference standard)} \\[3mm] D_T + \dfrac{A_T - D_T}{1 + e^{B_T[\log(\text{dose}_i) - C_T]}} + \epsilon_i, & x_i = 1 \text{ (i.e., test sample)}, \end{cases} \tag{7.4}$$

with parallel model:

$$y_i = D + \frac{A - D}{1 + e^{B[\log(\text{dose}_i) - C_S + (\Delta_C \times x_i)]}} + \epsilon_i. \tag{7.5}$$

Finally for the 5PL relationship, the unconstrained model is given by:

$$y_i = \begin{cases} D_S + \dfrac{A_S - D_S}{1 + e^{B_S[\log(\text{dose}_i) - C_S]^{E_S}}} + \epsilon_i, & x_i = 0 \text{ (i.e., reference standard)} \\[3mm] D_T + \dfrac{A_T - D_T}{1 + e^{B_T[\log(\text{dose}_i) - C_T]^{E_T}}} + \epsilon_i, & x_i = 1 \text{ (i.e., test sample)}, \end{cases} \tag{7.6}$$

with parallel model is given by:

$$y_i = D + \frac{A - D}{1 + e^{B[\log(\text{dose}_i) - C_S + (\Delta_C \times x_i)]^E}} + \epsilon_i. \tag{7.7}$$

Parallelism can be assessed by comparing the residuals under the parallelism constraint, as given in Equations (7.3, 7.5, 7.7), with the residuals for the corresponding unconstrained curves, given by Equations (7.2, 7.4, 7.6). The unconstrained models provide separate fits to each of the reference standard and test samples and will therefore offer a better overall fit to the data; that is, the sum of the squared residuals will be smaller for the unconstrained model. The parallelism F-test is then used to determine if the difference in the sum of squared errors between the constrained (SSE_{parallel}) and unconstrained (SSE_{separate}) models is statistically significant.

The F-statistic for parallelism is defined as:

$$F = \frac{(SSE_{\text{parallel}} - SSE_{\text{separate}})/(df_{\text{separate}} - df_{\text{parallel}})}{SSE_{\text{separate}}/df_{\text{separate}}},$$

and has degrees of freedom $df_1 = df_{\text{separate}} - df_{\text{parallel}}$ and $df_2 = df_{\text{separate}}$. The parallelism F-test is deemed significant at level α if F is greater than the critical value $F_{critical}$:

$$F > F_{critical} = F_{1-\alpha, df_1, df_2},$$

where $F_{1-\alpha, df_1, df_2}$ is the $(1 - \alpha)$ percentile of the F-distribution with df_1 and df_2 degrees of freedom. The degrees of freedom are explained below.

The numerator of the F-test statistic corresponds to the difference in the sum of squared errors between the separate and parallel models, divided by the change in the degrees of freedom. The change in the degrees of freedom corresponds to df_1 above and counts the number of parameters that are constrained to be equal in the parallel model. This will be $df_1 = 1$ (β_1) for a straight-line relationship; $df_1 = 3$ (A, B, D) for the 4PL relationship; and $df_1 = 4$ (A, B, D, E) for the 5PL.

The denominator of the F-test statistic corresponds to the mean squared error (MSE) for the unconstrained model and has degrees of freedom df_2. The degrees of freedom associated with the MSE again will depend on the relationship being fitted and will be given by $n - p$, where n is the total number of observations and p is the number of parameters required to fit the unconstrained dose-response relationship. For example, the straight-line relationship has two parameters (the slope and intercept), so for the two samples $p = 2 \times 2 = 4$ and $df_2 = n - 4$. For the 4PL relationship the number of parameters for the unconstrained fit will be $p = 8$ and $p = 10$ for the 5PL.

Figure 7.10 shows the separate 4PL fit in (a) and the parallel 4PL fit in (c) to the data in Table B.15. The corresponding residuals for these fits are in (b) and (d), respectively. From the separate fit, it is clear that the asymptotes for the reference standard and test sample are not equal. Therefore, the horizontal distance between the curves is not constant and the relative potency is not unique. The residuals for the separate lines model in (b) are smaller than the residuals for the parallel model in (d). The F-test for parallelism in this case is significant with a p-value < 0.001.

7.4.2 Significance tests for parallelism: binary data

A likelihood ratio test can be used to compare the constrained and unconstrained models [12].

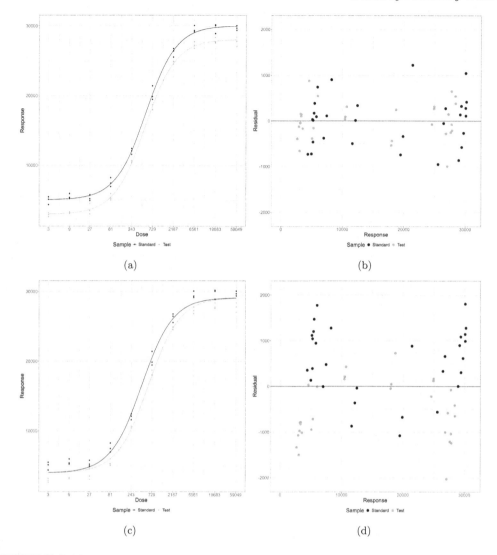

FIGURE 7.10

Separate 4PL fits to a reference standard and test sample in (a) and their residuals in (b). Parallel 4PL fit to the same data in (c) with the residuals to the parallel model fit in (d). There is a clear difference in the asymptotes of the standard and test samples. The residuals for the parallel fit are much larger than for the separate fits. The data for this plot can be found in Table B.15.

7.4.3   Equivalence tests for parallelism

Equivalence tests are the preferred method for assessing parallelism. For equivalence tests, firstly a metric must be chosen to measure the difference between the parameters for the reference standard and the test sample. Typically statistical software will calculate these metrics and their confidence intervals; we do not show their calculation here. There are many ways to define the metrics for comparing the parameters. Table 7.2 includes some of these. To provide an example, for an equivalence test for the slopes of a

TABLE 7.2

Possible metrics for parameter comparisons for parallelism tests.

Data type	Parameter	Difference	Ratio	Normalised difference
Continuous				
Straight-line	Slope	$\beta_{1T} - \beta_{1S}$	$\dfrac{\beta_{1T}}{\beta_{1S}}$	$\dfrac{(\beta_{1T} - \beta_{1S})}{(\beta_{1T} + \beta_{1S})/2}$
4PL/5PL	Slope parameter	$B_T - B_S$	$\dfrac{B_T}{B_S}$	$\dfrac{(B_T - B_S)}{(B_T + B_S)/2}$
	Left asymptote	$A_T - A_S$	$\dfrac{A_T}{A_S}$	$\dfrac{(A_T - A_S)}{(A_T + A_S)/2}$
	Right asymptote	$D_T - D_S$	$\dfrac{D_T}{D_S}$	$\dfrac{(D_T - D_S)}{(D_T + D_S)/2}$
	Range	$(D_T - A_T) - (D_S - A_S)$	$\dfrac{(D_T - A_T)}{(D_S - A_S)}$	
5PL	Asymmetry	$E_T - E_S$	$\dfrac{E_T}{E_S}$	$\dfrac{(E_T - E_S)}{(E_T + E_S)/2}$
Binary	Slope	$\beta_{1T} - \beta_{1S}$	$\dfrac{\beta_{1T}}{\beta_{1S}}$	$\dfrac{(\beta_{1T} - \beta_{1S})}{(\beta_{1T} + \beta_{1S})/2}$

straight-line relationship, the null and alternative hypotheses can be expressed in terms of a difference between the slopes:

$$H_0: |\beta_S - \beta_T| \geq \delta$$
$$H_A: |\beta_S - \beta_T| < \delta$$

or in terms of a ratio:

$$H_0: \frac{\beta_S}{\beta_T} \geq \delta \text{ or } \frac{\beta_S}{\beta_T} \leq \frac{1}{\delta}$$
$$H_A: \frac{1}{\delta} < \frac{\beta_S}{\beta_T} < \delta.$$

In many cases, there is no best choice of metric and any would be sufficient. In others, there is an obvious best choice. For example, if the estimate of an asymptote is close to zero, then using an equivalence test for the asymptote ratio between the reference and test samples can be risky. When the denominator approaches zero, the ratio can be infinite. This can result in a high number of failures even if the absolute difference between the asymptotes is negligible. Therefore, a metric based on the asymptote difference is often a safer choice.

Referring to Figure 7.10 and the data in Table B.15, the 90% confidence interval for the difference in the lower asymptotes is $(-2510, -1601)$; for the difference in the upper asymptotes is $(-2432, -1457)$; and for the difference in the slope parameters is $(-0.1342, 0.1068)$. When ensuring that the assumption of parallelism is met, all three of these confidence intervals must fall inside their respective equivalence bounds. Suppose the equivalence bounds for the difference of the lower asymptotes and the difference of the upper asymptotes were both $(-500, 500)$, and the equivalence bounds for the difference in the slope parameters was $(-0.2, 0.2)$. Then, this particular assay fails to demonstrate equivalence of the asymptotes but demonstrates equivalence of the slope parameters. Overall, however, parallelism has not been demonstrated. This result is consistent with the F-test for parallelism, but that will not always be the case.

7.5   Accuracy and precision of the potency estimate

The suitability tests discussed in the previous sections address whether the assay satisfies the statistical assumptions underlying the calculations. They do not directly address the accuracy or precision of the potency estimate.

Accuracy

If there is a QC sample in the bioassay, its estimated potency can be checked to ensure that it is in line with expectations. If not, the accuracy of the result for the test sample is not assured. The potency estimate for the QC sample, or a confidence interval for the potency, can be used as criteria.

If the QC sample is an independent replicate of the reference standard, then we could, for example, look to see if the CI for the RP falls within the bounds $(0.8, 1.25)$. These bounds should be chosen to be multiplicatively symmetric about 1.

Precision

The precision of the QC sample and test sample potency estimates can be directly assessed via their confidence intervals. The CI gives a range of plausible values for the potency (refer to Appendix A); the narrower the confidence interval the more precise the estimate. The ratio of the upper confidence limit to the lower confidence limit is a useful metric for precision. We define the precision factor, PF, as:

$$PF = \frac{\text{upper confidence limit}}{\text{lower confidence limit}}.$$

Usually, the 95% confidence interval is used.

7.6 Summary of suitability criteria

7.6.1 System suitability

The purpose of system suitability tests is to check that the bioassay is working correctly as a measurement system. For this, it is usual to examine the response of the bioassay with respect to samples with known properties: the reference standard and the QC sample (if used).

7.6.1.1 Reference standard

The reference standard is key to determining the assay result, as the test samples will be compared with it. If there is no QC sample included in the bioassay, then the only way to check that the system has performed correctly is to assess the behaviour of the reference standard. This can include the following:

1. Model fit;

2. Parameter estimates.

7.6.1.2 Quality control sample

A QC sample has known (or well estimated) relative potency. It should be biologically similar to the reference standard, thus, its dose-response curve should be parallel. If the reference standard and the QC sample are not parallel, then the bioassay cannot be assumed to be responding correctly. Checks for the QC sample can include the following:

1. Model fit;

2. Parallelism to the reference standard;

3. Value of the QC potency (estimate or confidence interval).

7.6.2 Sample suitability

To ensure that the RP is unique we need to ensure that the curve for the test sample is parallel to the reference standard. Tests for sample suitability only apply to that sample and failures do not necessitate that the entire assay be repeated. Checks for the test sample can include the following:

1. Model fit;

2. Parallelism to the reference standard;

3. PF for sample potency.

7.7 Chapter summary

- 'System suitability' tests check that the bioassay – the measurement system – performed appropriately in the current case, and 'sample suitability' tests check that each individual test sample behaved in a way to allow the measurement of its potency to be made.

- The two standard approaches to setting suitability criteria, significance testing and equivalence testing, are outlined and compared.

- Methods for setting limits for the suitability criteria in a range of scenarios are discussed, including the assessment of the fit of the dose-response relationship for the continuous data (linear, 4PL, 5PL cases) and binary data.

- Parallelism of the reference standard, QC and test samples is covered.

- Limiting the number of tests used to avoid unnecessary random failures is explained.

- Aspects of accuracy and precision for the bioassay are discussed and ways of assessing these aspects directly are described.

Part III

Validation, monitoring and modifications

8

Qualification and validation of bioassays

Evaluation of performance continues throughout the lifecycle of a bioassay. During development, performance is assessed in a series of development and qualification studies. For filing a licence application, a biological product should have a bioassay method that has been formally validated, i.e., demonstrated to have acceptable performance [20, 21]. 'Acceptable performance' usually relates to the release procedure and therefore involves the release procedure format.

During validation studies, the method documentation (or standard operating procedure) should be followed. The documentation should include ranges for robustness parameters such as temperature, incubation time, etc. Ruggedness factors (factors which are qualitative in nature and are likely to change over time, such as analyst, reagent lot, etc.) should be deliberately varied during validation in order to capture the extent of the variability and to estimate the assay performance realistically.

Following validation, the bioassay should be monitored to ensure that the performance continues to be acceptable, and corrective action must be taken where necessary. This process is sometimes referred to as continuous performance verification or ongoing procedure performance verification (OPPV)[63].

8.1 Measuring bioassay performance

As discussed in Chapter 2, performance characteristics should be prospectively selected to allow evaluation of the bioassay procedure's fitness for its intended use: essentially, it must have a high chance of producing a result that leads to the correct decision, for example whether or not the tested lot may be released. The analytical target profile (ATP) should capture the performance characteristics for a bioassay procedure with associated performance criteria.

DOI: 10.1201/9781003449195-8

TABLE 8.1

Bioassay performance characteristics.

Per true potency level	Across tested potency levels
• Accuracy	• Accuracy / linearity
• Intermediate precision	• Intermediate precision
• Tolerance interval: reportable value	• Tolerance interval: reportable value for manufacturing process
	• Range

8.1.1 Bioassay performance characteristics

This chapter will focus on the performance characteristics listed in Table 8.1. These characteristics are defined in Chapter 2. Traditionally, accuracy, intermediate precision, linearity and range have been the focus of bioassay validation. More recently, attention has been given to developing combined measures of accuracy and precision as potential replacements for the separate assessment of accuracy and precision. Here we consider a tolerance interval for the reportable value (RV).

In a validation study, selected characteristics are validated in the context of the release procedure. That is, conformance of the assay method to the stated acceptance criteria, coupled with the release procedure format (k_F runs of the method in each of n_F sessions), demonstrates that the bioassay method is fit for use in the release procedure. However, if studies are carefully designed, more general conclusions can be drawn about the performance of the bioassay in other contexts/for other procedures such as qualification of a new reference standard or transfer of the bioassay to a new facility (Chapter 10).

8.1.2 Bioassay qualification

We will use the term 'qualification' to mean the evaluation of performance characteristics in the absence of specific target values for these. The result of a qualification study is a set of values for the characteristics that the bioassay seems capable of achieving. These values can be considered in the light of the analytical target profile for the product, to inform the format (or replication strategy) for the release procedure and other procedures such as stability testing. They can also inform the design of the validation study.

8.1.3 Bioassay validation

Bioassay validation is similar to bioassay qualification but requires acceptance criteria to be set before the start of the study for the performance characteristics possibly defined by the ATP. The goal of the validation study is to prove that the assay has met its acceptance criteria and that it is fit for purpose.

8.1.3.1 Co-validation and partial validation

We do not cover co-validation and partial validation specifically, but we include descriptions for completeness. Chapter 10 also contains additional relevant details.

When a method is to be transferred from a development laboratory to another laboratory, it is possible to perform a validation study involving both laboratories. This process is known as co-validation [61]. Co-validation can accelerate the final validation of the method, but involves close collaboration between the two laboratories and needs careful planning [57].

Partial validation has been defined in [5] as the demonstration of assay reliability following a modification of an existing bioanalytical method that has previously been fully validated. The nature of the modification will determine the extent of validation required.

8.1.4   Bioassay qualification and validation studies

Evaluation of the performance of a bioassay involves analysis of the results across repeated runs and sessions for the same test sample, usually with a known true value to allow assessment of accuracy, as described in Chapters 2 and 3. During these evaluations, the assay format that will be used for the release procedure can be decided based on the performance that is required.

This chapter is presented in terms of validation studies for relative potencies. However, the methods apply to any result, whether relative or absolute. For qualification studies, where there is no penalty associated with wide confidence intervals, the study size may be smaller, but the principles still apply.

8.2   Performance criteria

To define the aims of a validation study, the first step is to identify performance characteristics and their validation acceptance criteria, guided by the ATP. Bioassay acceptance criteria are usually set separately for accuracy and intermediate precision, and sometimes linearity. Alternatively, accuracy (systematic error) and precision (random error) can be combined into a single measure that captures 'total analytical error'.

Assessment of whether parameter values meet their criteria should be based on confidence limits for the parameters [27, 59]. For successful validation it is important that the bioassay method is expected to achieve better accuracy and precision than required by the criteria. Basing the assessment on the point estimates of the criteria instead of the confidence limits will result in a higher chance of success in the study (given the same limits). However, it is likely to lead to overly optimistic conclusions regarding the accuracy and precision and could lead to higher-than-expected failures for batch release.

8.2.1   Accuracy and intermediate precision

Recall from Chapter 2, accuracy is defined via relative bias as:

$$RB = 100 \times \frac{[\text{antilog}\,(\mu_{\log RP}) - TRP]}{TRP}\%,$$

where $\mu_{\log RP}$ is the mean value of the log-transformed relative potencies and TRP is the true potency of the batch.

Intermediate precision, IP, is defined as the geometric coefficient of variation for repeated runs of the bioassay method:

$$IP = 100 \times \left[\text{antilog} \left(\sqrt{\sigma^2_{\log RP}} \right) - 1 \right] \%.$$

The variance of the log-transformed relative potencies, denoted by $\sigma^2_{\log RP}$, is a combination of variance components for the method. The components relate to the way in which the execution of the method is repeated in the release format:

$$\sigma^2_{\log RP} = \sigma^2_{\text{Between sessions}} + \sigma^2_{\text{Within session}}.$$

Therefore, the IP for repeated runs of the bioassay method is:

$$IP = 100 \times \left[\text{antilog} \left(\sqrt{\sigma^2_{\text{Between sessions}} + \sigma^2_{\text{Within session}}} \right) - 1 \right] \%.$$

8.2.2 Acceptance criteria for accuracy and precision

The bioassay release procedure will have been designed to produce a reportable value that has an acceptable chance of falling in specification (see Section 3.4.3). The estimates of accuracy and precision used for this process should have been conservative. For example, see Table 3.8. If the accuracy is RB = -16.7% and the intermediate precision is IP = 20%, then a release format of 8 sessions, with 1 method run each, provides an out-of-specification (OOS) proportion of 0.8%. These calculations assume that the specification limits are $(0.70, 1.43)$ and the manufacturing process variance is assumed to be centred at RP = 1 and to account for one fifth (0.2) of the reportable value variance. Assuming that 1% is an acceptable OOS proportion, validation study acceptance limits of $(-17\%, 20\%)$ for RB and 20% for IP could be reasonable as the ATP for a release bioassay consisting of 8 sessions, with 1 method run each (i.e., $n_F = 8$, $k_F = 1$).

Figure 8.1 illustrates the distribution of true values of RP from the manufacturing process (in grey), overlaid with the distribution of reportable values from the release procedure (in black). The sum of the areas under the black curve to the left of the lower specification limit and to the right of the upper specification limit represents the OOS probability for lots from the manufacturing process.

Assuming the release assay format has been designed conservatively, allowing for more bias and imprecision than is estimated to exist, the validation study can be designed such that there is a good chance that the confidence intervals for the parameters will lie within the limits. For the assay evaluated in Chapters 2 and 3 the estimates of RB and IP were -10.8% and 13.4%, respectively; therefore, the above format of 8 runs would be conservative.

The criterion for accuracy would be that the (two-sided) 90% confidence interval for RB falls within $(-17\%, 20\%)$. The criterion for intermediate precision would be that the upper (one-sided) 95% confidence limit for IP falls below 20%. The closer the true values

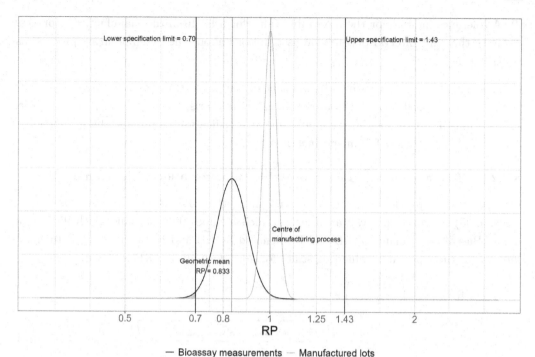

— Bioassay measurements — Manufactured lots

FIGURE 8.1

Distributions of true values of RP from the manufacturing process and reportable values from the release procedure.

are to the acceptance limits, the bigger the study required to have a good chance of passing the criteria. The number of sessions and method runs required in the study is covered in Section 8.3.4.

8.2.2.1 Combinations of accuracy and precision

In Chapter 2, we introduced a range of measures which combine accuracy and precision. We could consider the proportion of OOS results. However, the calculations of the estimate of this quantity and its confidence limit are complex when there is more than one component of variance to be handled. Therefore, for the purpose of validation, we will focus on a tolerance interval for the reportable value. For simplicity, we will restrict the discussion to the case, where $k_F = 1$: that is, the assay format for the reportable value is n_F sessions, each including one run of the bioassay method.

Recall that a $(p,\ 1-\alpha)$ tolerance interval (TI) includes at a proportion p of the population with confidence $1 - \alpha$. We will consider two approaches:

1. Tolerance intervals for the reportable value for lots with known potency.

   - These provide an assessment of the capability of the assay to give a correct decision for a lot. The tolerance interval should lie outside the specification limits for a lot with true value which is too low or too high for safety or efficacy.

2. A tolerance interval for the reportable value for a lot from the manufacturing process. Note that this interval depends on assumptions about the manufacturing process distribution.

- This provides an indirect assessment of the proportion, p, of lots expected to fall out of specification overall: the manufacturer's risk. The value of p can be varied until the tolerance interval falls within specification to provide the proportion of lots expected to fall in specification.

8.2.2.2 Tolerance interval for reportable value for a lot with given TRP

Consistent with previous requirements (no more than 1% OOS), we can construct a (0.99, 0.95) tolerance interval: an interval which includes 99% of values with 95% confidence. Based on estimates of RB and IP, denoted by $\widehat{\text{RB}}$ and $\widehat{\text{IP}}$, a (0.99, 0.95) tolerance interval for the reportable value (log scale) for a lot with given TRP is given by:

$$\log\left(\text{TRP}\right) + \log\left(\frac{\widehat{\text{RB}}}{100} + 1\right) \pm$$

$$z_{0.995} \times \sqrt{\frac{\left[(\widehat{\sigma}^2_{\text{Between sessions}} + \widehat{\sigma}^2_{\text{Within session}})/n_F\right]\left(1 + \frac{1}{n}\right) \times df}{\chi^2_{0.05,\,df}}}. \tag{8.1}$$

Note that n is the sample size for the estimation of RB and df are the degrees of freedom for the estimation of the variance [7]. See Section 2.5.5 where the formula is given for a format with one session, i.e., $n_F = 1$ and so $df = n - 1$.

The acceptance criterion for the TI would be that it falls within the (log-transformed) specification limits, $[\log(\text{LSL}), \log(\text{USL})]$. This is consistent with 99% of lots with true potency of TRP falling in specification (with 95% confidence).

8.2.2.3 Tolerance interval for reportable value for a lot from the manufacturing distribution

Similar to the above, a (0.99, 0.95) tolerance interval for the reportable value is given by:

$$\mu_{\text{Product}} + \log\left(\frac{\widehat{\text{RB}}}{100} + 1\right) \pm$$

$$z_{0.995} \times \sqrt{\frac{\left[(\widehat{\sigma}^2_{\text{Between sessions}} + \widehat{\sigma}^2_{\text{Within session}})/n_F\right] \times \left(1 + \frac{1}{n}\right) \times df}{(1 - P) \times \chi^2_{0.05,\,df}}}, \tag{8.2}$$

where μ_{Product} is the assumed mean value (log scale) for the manufacturing process and P is the assumed proportion of the variance accounted for by the manufacturing process.

8.2.3 Linearity

When linearity can be demonstrated, the bias is constant across the TRP and a pooled estimate of the bias can be reported. The linear relationship between observed and true relative potencies can be expressed as:

$$\log(\text{RP}) = \beta_0 + \beta_1 \times \log(\text{TRP}) + \epsilon, \tag{8.3}$$

where ϵ is a normal error term with mean 0 and variance $\sigma^2_{\log \mathrm{RP}}$, then:

$$\mu_{\log \mathrm{RP}} = \beta_0 + \beta_1 \times \log(\mathrm{TRP})$$

and

$$\mathrm{RB} = 100 \times \{\mathrm{antilog}\left[\beta_0 + (\beta_1 - 1) \times \log(\mathrm{TRP})\right] - 1\}.$$

If $\beta_1 = 1$ then RB is constant across all values of TRP and the assay is said to achieve linearity. Acceptance criteria for linearity are defined in terms of the slope, β_1, of the linear relationship between $\log(\mathrm{RP})$ and $\log(\mathrm{TRP})$ defined in Equation (8.3). A typical criterion would be that the two-sided 90% confidence interval for β_1 falls within (0.80, 1.25). When linearity is not demonstrated, it is common to drop dose groups from the ends of the range until a (continuous) set of doses which pass the linearity test is identified.

When the number of measurements per session for a TRP in the study is greater than 1, to avoid unnecessary complexity resulting from variance components, the linear model may be fitted to the session mean per TRP. (If the individual replicate measurements are to be used instead, a mixed model should be used to ensure that the confidence interval is estimated appropriately.)

8.2.4 Range

The range is the set of potency levels for which the bioassay has been demonstrated to have acceptable performance. The range of the bioassay should ideally include the interval between the lower specification limit and the upper specification limit. When linearity is a condition for acceptable performance of the assay, this implies that the relative bias is constant across the range.

8.3 Validation study protocol

The validation study protocol should detail the following:

- Performance characteristics to be tested and their acceptance criteria;

- Test samples/target potency levels to be included;

- Conditions for the execution of the method;

- Number of study sessions, n_s, and number of method runs per study session, k_s.

- Analysis methodology to evaluate the performance characteristics.

The study protocol should detail how the statistical analysis will be conducted.

8.3.1 Performance characteristics

The performance characteristics and their acceptance criteria have been discussed in Section 8.2. These need to be specified prospectively in the protocol so that the performance of the assay can be assessed objectively.

8.3.2   Test samples

The assessment of accuracy requires test samples with a known relative potency. We refer to this known potency as the TRP. The TRP values need to cover the required range for the bioassay. This should be at least as wide as the specification.

Test samples with known relative potency are typically made from dilutions of the reference standard, so that their true potency is known, by definition. A study using such samples is called a dilutional linearity study. The TRP values should be equally spaced on the log scale. For example:

$$\{0.50, \ 0.71, \ 1.00, \ 1.41, \ 2.00\}.$$

Test samples whose true potency is unknown can be included in the study for the assessment of precision but not of accuracy.

8.3.3   Design factors

The study should include variation in levels of all ruggedness factors that could affect the result, so that the estimated precision is not overly optimistic. Therefore, an appropriate number of operators should be included, critical reagent batches should be varied if possible and the study should be conducted over a period of time that is likely to capture representative variability of the assay measurement. Design of experiments techniques can be beneficial in the planning of the validation study.

8.3.4   Number of study runs

If it is feasible for the bioassay method to include more than one run within a session, then the validation study should be designed to evaluate both 'within session' and 'between session' components of variance. This can be useful even if the release procedure will, in fact, include only one run per session because it can inform the design of other procedures (such as stability testing). The simplest approach is to match the release procedure in the validation study design. However, it is not necessary to replicate the release procedure in the study in terms of the numbers of method runs conducted per session [59]. For the case where $n_F = 2$, $k_F = 3$, for example, depending on resources and the desire to represent a range of ruggedness factors, it might be preferable to design the study with two runs of the method per session. The components of variance can be used to estimate the intermediate precision using mathematical techniques.

For assays where only a single method run is feasible within a session, k_F will be 1 and the validation study sessions will also include only one method run. In this case the within session variance component cannot be estimated (and is not required).

Success for the validation study is achieved when the confidence limits for both the RB and the IP (or the single combined criterion, if used) sit within their acceptance criteria. The number of study runs required is driven by the following:

1. The required chance of study success, typically 90%;

2. The true IP for the bioassay;

3. The true RB for the bioassay;

4. Acceptance limits for the performance characteristics.

For each level tested, we denote the number of study sessions by n_s and the number of method runs within each study session as k_s.

Conventionally, the sample size is applied to each individual true potency level. We note that when linearity holds, the estimate of the relative bias should be pooled across all levels to give a single estimate of accuracy. Similarly, the estimate of the intermediate precision should be pooled across all levels when appropriate. Therefore, the methods below for calculating the sample size are likely to be conservative.

8.3.4.1 Sample size based on accuracy

A simple approach to the sample size is to focus on the accuracy, and to consider only n_s. When $k_s > 1$ this will be a conservative approach. We assume that for the validation study to be successful, the 90% confidence interval for the RB must fall within pre-specified (symmetric) acceptance limits. Let RB* be the positive (upper) validation acceptance limit for the RB. Then, the symmetric acceptance limits will be given by:

$$\left[\frac{-\text{RB}^*}{(1 + \frac{\text{RB}^*}{100})}, \text{RB}^*\right].$$

When calculating the sample size, assumptions about the true RB and IP must be made. The choice of these parameters should be informed by historical data. There are several available formulae for the required sample size [6]. The following provides a minimum value for n_s, assuming that the true RB is zero and the true intermediate precision is IP:

$$n_s \geq \frac{\left(t_{0.95,n_s-1} + t_{(1-\frac{\beta}{2}),n_s-1}\right)^2 \times \left[\log\left(1 + \frac{\text{IP}}{100}\right)\right]^2}{\left[\log\left(1 + \frac{\text{RB}^*}{100}\right)\right]^2},$$

where $t_{p,\,df}$ is the $100p$ percentile of the t distribution with df degrees of freedom, and $(1 - \beta)$ is the chance of success of the study with regard to the validation of accuracy. If the true relative bias, RB, is assumed to be positive, then the formula becomes:

$$n_s \geq \frac{\left(t_{0.95,n_s-1} + t_{(1-\frac{\beta}{2}),n_s-1}\right)^2 \times \left[\log\left(1 + \frac{\text{IP}}{100}\right)\right]^2}{\left[\log\left(1 + \frac{\text{RB}^*}{100}\right) - \log\left(1 + \frac{\text{RB}}{100}\right)\right]^2}.$$

If the true relative bias, RB, is assumed to be negative, then the formula becomes:

$$n_s \geq \frac{\left(t_{0.95,n_s-1} + t_{(1-\frac{\beta}{2}),n_s-1}\right)^2 \times \left[\log\left(1 + \frac{\text{IP}}{100}\right)\right]^2}{\left[\log\left(1 + \frac{\text{RB}^*}{100}\right) + \log\left(1 + \frac{\text{RB}}{100}\right)\right]^2}.$$

Note that the sample size appears on both sides of the formula. It is not possible to solve explicitly so an iterative approach must be used.

Table 8.2 provides some examples of the sample sizes required for various combinations of RB and IP assuming probability of success is $(1 - \beta) = 0.9$.

TABLE 8.2

Numbers of study runs required based on the criterion for the relative bias for a chance of validation success of 0.9 for relative bias.

RB* Limits	True RB	True IP 14%	10%	8%
	−11%	45	25	17
(−17%, 20%)	3%	10	7	5
	0%	8	5	4

8.3.4.2 Sample size based on precision

Alternatively/additionally, the validation study can be based on the required sample size to meet the precision acceptance criterion. Again, we consider estimating the number of sessions, n_s. Suppose the criterion for precision is that the upper 95% (one-sided) confidence limit for IP falls below IP*. Equivalently, on the log scale, the upper 95% (one-sided) confidence limit for the variance, $\sigma^2 = \left[\log\left(\frac{IP}{100} + 1\right)\right]^2$, falls below $\left[\log\left(\frac{IP^*}{100} + 1\right)\right]^2$. We want the probability of this event to be high: say at least 0.9.

Let $(l = 1, \cdots, L)$ index the target (true) relative potency values for the validation study. Then at each level individually, the upper 95% (one-sided) confidence limit for the variance is given by:

$$\frac{s_l^2 \times (n_s - 1)}{\chi^2_{0.05,(n_s-1)}},$$

where s_l^2 is the estimate of the variance estimate for level l and $\chi^2_{0.05,n_s-1}$ is the 5th percentile of the chi-square distribution with $n_s - 1$ degrees of freedom. The distribution of $\frac{s_l^2 \times (n_s-1)}{\sigma^2}$ is chi-squared with $(n_s - 1)$ degrees of freedom. Therefore, the probability that the upper 95% (one-sided) confidence limit for the variance, σ^2, is less than $\left[\log\left(\frac{IP^*}{100} + 1\right)\right]^2$ is given by:

$$Prob\left\{\frac{s_l^2 \times (n_s - 1)}{\chi^2_{0.05,(n_s-1)}} < \left[\log\left(\frac{IP^*}{100} + 1\right)\right]^2\right\}$$

$$= Prob\left\{\frac{s_l^2 \times (n_s - 1)}{\sigma^2} < \frac{\left[\log\left(\frac{IP^*}{100} + 1\right)\right]^2 \times \chi^2_{0.05,(n_s-1)}}{\sigma^2}\right\}.$$

For this probability to be at least 0.9,

$$Prob\left\{\chi^2_{(n_s-1)} < \frac{\left[\log\left(\frac{IP^*}{100} + 1\right)\right]^2 \times \chi^2_{0.05,(n_s-1)}}{\left[\log\left(\frac{IP}{100} + 1\right)\right]^2}\right\} \geq 0.9.$$

The value of n_s needs to be found iteratively.

On an individual level basis, the sample sizes required tend to be infeasibly high unless there is a big gap between the limit, IP*, and the assumed IP. Therefore, ideally, the IP

TABLE 8.3

Numbers of study runs required for a chance of validation success of 0.9 for intermediate precision: estimate per level, or pooled over 5 TRP levels.

IP* Limit	True IP	n_s per level, not pooled across levels	n_s per level, pooled across 5 potency levels
20%	14%	43	10
20%	10%	13	4
20%	8%	8	3
12%	8%	32	8

and its confidence interval would be estimated based on the levels pooled together. In that case, approximately,

$$Prob\left\{\chi^2_{L\cdot(n_s-1)} < \frac{\left[\log\left(\frac{\text{IP}^*}{100}+1\right)\right]^2 \times \chi^2_{0.05,L\cdot(n_s-1)}}{\left[\log\left(\frac{\text{IP}}{100}+1\right)\right]^2}\right\} \geq 0.9. \quad (8.4)$$

Table 8.3 provides numbers of study runs required for a chance of validation success of 0.9 for intermediate precision: estimate per level, or pooled over 5 TRP levels.

8.3.4.3 Sample size based on tolerance interval

Unlike for accuracy and precision, estimating the sample size for a tolerance interval criterion is complex. Simulation is required [7].

8.3.5 Calculation of RB, IP using validation study data

The protocol needs to state the methods for analysis of each performance criterion; we detail these here. Let $(l = 1, \cdots, L)$ index the target (true) relative potency values for the validation study. Then let $\widehat{\log \text{RP}}_{ijl}$ be the log relative potency estimate for the jth method run, $j = 1, \ldots, k_s$, of the ith study session, $i = 1, \ldots, n_s$, at level l.

As for the analyses in Chapter 2, all analysis for the estimation of bias and precision is conducted on the log scale. We first define the some quantities needed for the analysis. The mean log RP, at level l, within study session i, can be found by averaging over the k_s method runs:

$$\widehat{\log \text{RP}}_{i\cdot l} = \frac{\sum_{j=1}^{k_s} \widehat{\log \text{RP}}_{ijl}}{k_s}.$$

The mean log RP, for level l, can be estimated by averaging the across the $k_s \times n_s$ results at that level:

$$\widehat{\log \text{RP}}_{\cdot\cdot l} = \frac{\sum_{i=1}^{n_s} \sum_{j=1}^{k_s} \widehat{\log \text{RP}}_{ijl}}{k_s \times n_s}.$$

Finally, the mean log RP across all levels can be calculated by averaging over all results as:

$$\widehat{\log \text{RP}}_{\cdots} = \frac{\sum_{l=1}^{L} \sum_{i=1}^{n_s} \sum_{j=1}^{k_s} \widehat{\log \text{RP}}_{ijl}}{L \times k_s \times n_s}.$$

8.3.5.1   Estimation of RB: each level separately

To estimate the accuracy, we begin with estimating the relative potency at TRP_l as:

$$\widehat{RP}_l = \text{antilog}\left(\widehat{\log RP}_{..l}\right),$$

and the corresponding estimate of RB_l is given by:

$$\widehat{RB}_l = 100 \times \left(\frac{\widehat{RP}_l}{\text{TRP}_l} - 1\right)\%.$$

For a confidence interval for RB_l, we work with the mean values per session; it is unnecessary for this purpose to separate the within and between session components of the variability. The standard deviation (log scale) for level l is estimated as:

$$s_l = \sqrt{\frac{\sum_{i=1}^{n_s}\left(\widehat{\log RP}_{i\cdot l} - \widehat{\log RP}_{..l}\right)^2}{(n_s - 1)}}. \tag{8.5}$$

This provides a direct estimate, for level l, of $\left(\sigma^2_{\text{Between sessions}} + \frac{\sigma^2_{\text{Within session}}}{k_s}\right)$.

A two-sided 90% confidence interval, $\left(\widehat{RP}_{l,LCL}, \widehat{RP}_{l,UCL}\right)$, for RP_l is given by:

$$\widehat{RP}_{l,LCL} = \text{antilog}\left(\widehat{\log RP}_{..l} - t_{0.95,n_s-1}\frac{s_l}{\sqrt{n_s}}\right),$$

$$\widehat{RP}_{l,UCL} = \text{antilog}\left(\widehat{\log RP}_{..l} + t_{0.95,n_s-1}\frac{s_l}{\sqrt{n_s}}\right).$$

A two-sided 90% confidence interval for RB_l is given by:

$$\left[100 \times \left(\frac{\widehat{RP}_{l,LCL}}{\text{TRP}_l} - 1\right)\%,\; 100 \times \left(\frac{\widehat{RP}_{l,UCL}}{\text{TRP}_l} - 1\right)\%\right].$$

8.3.5.2   Estimation of RB: pooled standard deviation across levels

In the previous section, only the data for an individual level were used to estimate the confidence interval for the RB. However, if it is justifiable to pool the levels to get a common estimate of the variance, this can result in a more precise estimate, i.e., a narrower confidence interval. The L estimates of the standard deviation (log scale) can be compared using Bartlett's test [1]. If the difference amongst the estimates is not significant (at, for example, the 5% level) then a pooled estimate of the standard deviation can be calculated as:

$$s = \sqrt{\frac{\sum_{l=1}^{L}\sum_{i=1}^{n_s}\left(\widehat{\log RP}_{i\cdot l} - \widehat{\log RP}_{..l}\right)^2}{L(n_s - 1)}}.$$

The test statistic for Bartlett's test in this (balanced) case is:

$$\chi^2 = \frac{(n_s - 1)^2\left[L \cdot \log(s^2) - \sum_{l=1}^{L}\log(s_l^2)\right]}{(n_s - 1) + (L+1)/3L},$$

which has an approximately χ^2_{L-1} distribution. Thus if $\chi^2 < \chi^2_{0.95,\ L-1}$ then it is acceptable to pool the estimates of standard deviation and to use this estimate, s, for the confidence intervals for RB.

For a two-sided 90% confidence interval, $\left(\widehat{RP}_{l,LCL}, \widehat{RP}_{l,UCL}\right)$, for RP_l based on the pooled estimate of the standard deviation, s:

$$\widehat{RP}_{l,LCL} = \text{antilog}\left(\widehat{\log RP}_{..l} - t_{0.95,L(n_s-1)}\frac{s}{\sqrt{n_s}}\right),$$

$$\widehat{RP}_{l,LCU} = \text{antilog}\left(\widehat{\log RP}_{..l} + t_{0.95,L(n_s-1)}\frac{s}{\sqrt{n_s}}\right).$$

For a two-sided 90% confidence interval for RB_l based on the pooled estimate of the standard deviation, s:

$$\left[100 \times \left(\frac{\widehat{RP}_{l,LCL}}{\text{TRP}_l} - 1\right)\%,\ 100 \times \left(\frac{\widehat{RP}_{l,UCL}}{\text{TRP}_l} - 1\right)\%\right].$$

8.3.5.3 Linearity and pooled estimate of RB across TRP levels

Linearity can be assessed on the log-log scale, comparing $\log(\text{RP})$ and $\log(\text{TRP})$. The straight-line relationship is expressed as (see Section 2.3.2):

$$\mu_{\log \text{RP}} = \beta_0 + \beta_1 \times \log(\text{TRP}).$$

This is equivalent to:

$$\log\left(\frac{\text{RB}}{100} + 1\right) = \beta_0 + (\beta_1 - 1) \times \log(\text{TRP}).$$

Therefore, if $\beta_1 = 1$, RB is constant across all values of TRP, at $100 \times (\text{antilog}(\beta_0) - 1)$.

Assuming the standard deviation is constant across the levels, a simple linear regression can be fitted to the data to provide estimates and confidence intervals for β_0 and β_1. If the confidence interval for β_1 falls within a pre-stated interval around 1 then we conclude that β_1 is equivalent to 1 and the RB does not vary significantly across the levels tested. When this occurs, then it is appropriate to calculate a single pooled estimate of RB as follows.

Referring to Equation (2.4), per level, l, we have:

$$\log\left(\frac{\text{RB}_l}{100} + 1\right) = \log\left(\text{RP}_l\right) - \log\left(\text{TRP}_l\right).$$

Averaging this difference between the logged values of the measured and true relative potencies across all levels provides a pooled estimate of the RB on the log scale. A pooled estimate of RB is given by:

$$\widehat{\text{RB}} = 100 \times \left[\frac{\text{antilog}\left(\widehat{\log \text{RP}}_{...}\right)}{\text{antilog}\left(\overline{\log \text{TRP}}_{...}\right)} - 1\right],$$

TABLE 8.4

Variance components table for $\widehat{\text{IP}}_l$. DF = degrees of freedom, MSS = mean sum of squares and $\mathbb{E}[\text{MSS}]$ = expected value of mean sum of squares.

Source	DF	MSS	$\mathbb{E}[\text{MSS}]$
Between sessions	$n_s - 1$	$\text{MSS}_{\text{Between}}$	$k_s \cdot \sigma^2_{\text{Between session}}$ $+\sigma^2_{\text{Within session}}$
Within session	$n_s(k_s - 1)$	$\text{MSS}_{\text{Within}}$	$\sigma^2_{\text{Within session}}$

where the denominator in the fraction of the above equation denotes the mean value of log(TRP) across all observations:

$$\overline{\log(\text{TRP})}_{...} = \frac{\sum_{l=1}^{L}\sum_{i=1}^{n_s}\sum_{j=1}^{k_s}\log(\text{TRP}_{ijl})}{L \times k_s \times n_s}.$$

For a study where the potency levels are symmetrically balanced on either side of 1 and all samples are tested an equal number of times, $\overline{\log(\text{TRP})}_{...} = 0$.

For a two-sided 90% confidence interval for the pooled RB:

$$100 \times \left(\frac{\text{antilog}\left(\widehat{\log \text{RP}}_{...} - t_{0.95, L(n_s-1)}\frac{s}{\sqrt{Ln_s}}\right)}{\text{antilog}\left(\overline{\log \text{TRP}}_{...}\right)} - 1 \right) \%,$$

$$100 \times \left(\frac{\text{antilog}\left(\widehat{\log \text{RP}}_{...} + t_{0.95, L(n_s-1)}\frac{s}{\sqrt{Ln_s}}\right)}{\text{antilog}\left(\overline{\log \text{TRP}}_{...}\right)} - 1 \right) \%.$$

8.3.5.4   Estimation of IP: each level separately

For IP, the two separate components of variance must be taken into account. Statistical software such as R [49] and the VCA [54] package are available. Firstly, IP is estimated at each level separately. The estimates are compared and, if appropriate, they can be pooled across the levels. Table 8.4 shows the analysis of variance table for estimation of the variance components at each level, separately. For the formulae for the expected mean squares refer to [8]. The estimates of the variance components are:

- Within session variance:

$$\hat{\sigma}^2_{\text{Within session}} = \text{MSS}_{\text{Within}};$$

- Between session variance:

$$\hat{\sigma}^2_{\text{Between sessions}} = \frac{(\text{MSS}_{\text{Between}} - \text{MSS}_{\text{Within}})}{k_s};$$

- Total variance:

$$\hat{\sigma}^2_{\text{Between sessions}} + \hat{\sigma}^2_{\text{Within session}} = \text{MSS}_{\text{Between}}\frac{1}{k_s} + \text{MSS}_{\text{Within}}\left(1 - \frac{1}{k_s}\right).$$

The estimate of intermediate precision at level l is then given by:

$$\widehat{\text{IP}}_l = 100 \times \left\{ \text{antilog} \left[\sqrt{\text{MSS}_{\text{Between}} \frac{1}{k_s} + \text{MSS}_{\text{Within}} \left(1 - \frac{1}{k_s}\right)} \right] - 1 \right\} \%.$$

The upper confidence limit for $\widehat{\text{IP}}_l$ can be found by following Section 3.4.2.

One method run per session

When $k_s = 1$, the standard deviations defined in Equation (8.5) provide estimates of the intermediate precision. At level l:

$$s_l^2 = \frac{\sum_{i=1}^{n_s} \left(\widehat{\log \text{RP}}_{i \cdot l} - \widehat{\log \text{RP}}_{\cdot \cdot l}\right)^2}{(n_s - 1)}$$

$$= \hat{\sigma}_{\text{Between sessions}}^2 + \hat{\sigma}_{\text{Within session}}^2.$$

Hence

$$\widehat{\text{IP}}_l = 100 \times [\text{antilog}\,(s_l) - 1]\,\%, \tag{8.6}$$

with an upper one-sided 95% confidence limit:

$$100 \times \left\{ \text{antilog} \left(s_l \sqrt{\frac{(n_s - 1)}{\chi_{0.05, n_s - 1}^2}} \right) - 1 \right\} \%, \tag{8.7}$$

where $\chi_{0.05, n_s - 1}^2$ is the 5th percentile of the chi-square distribution with $n - 1$ degrees of freedom.

8.3.5.5 Estimation of IP: pooled across levels

For the comparison of the estimates across the levels, although Bartlett's test cannot be applied directly (as it does not handle components of variance), it can be applied to the $L \times n_s$ within session estimates at each level:

$$s_{il} = \sqrt{\frac{\sum_{j=1}^{k_s} \left(\widehat{\log \text{RP}}_{ijl} - \widehat{\log \text{RP}}_{i \cdot l}\right)^2}{(k_s - 1)}}.$$

A pooled estimate of the standard deviation is calculated as:

$$s = \sqrt{\frac{\sum_{l=1}^{L} \sum_{i=1}^{n_s} \sum_{j=1}^{k_s} \left(\widehat{\log \text{RP}}_{ijl} - \widehat{\log \text{RP}}_{i \cdot l}\right)^2}{L \cdot n_s (k_s - 1)}}.$$

The test statistic for Bartlett's test in this (balanced) case is:

$$\chi^2 = \frac{(k_s - 1)^2 \left[L \cdot n_s \cdot \log(s^2) - \sum_{l=1}^{L} \sum_{i=1}^{n_s} \log(s_{il}^2) \right]}{(k_s - 1) + [(L \cdot n_s) + 1]/(3L \cdot n_s)}$$

which has approximately a $\chi_{(L \times n_s)-1}^2$ distribution. Thus if $\chi^2 < \chi_{0.95,\,(L \times n_s)-1}^2$ then it is acceptable to pool the estimates of within session standard deviation. If, in addition, the

TABLE 8.5

Variance components calculations for $\widehat{\text{IP}}$, where an interaction is assumed to exist between levels and sessions. DF = degrees of freedom, MSS = mean sum of squares and $\mathbb{E}[\text{MSS}]$ = expected value of mean sum of squares.

Source	DF	MSS	$\mathbb{E}[\text{MSS}]$
Between levels	$L-1$	$\text{MSS}_{\text{Levels}}$	Not relevant – fixed effect
Between sessions	$n_s - 1$	$\text{MSS}_{\text{Between}}$	$L \cdot k_s \cdot \sigma^2_{\text{Between session}} +$ $k_s \cdot \sigma^2_{\text{interaction}} + \sigma^2_{\text{Within session}}$
Levels × sessions	$(L-1) \cdot (n_s - 1)$	$\text{MSS}_{\text{Interaction}}$	$k_s \cdot \sigma^2_{\text{Interaction}} + \sigma^2_{\text{Within session}}$
Within session	$L \cdot n_s \cdot (k_s - 1)$	$\text{MSS}_{\text{Within}}$	$\sigma^2_{\text{Within session}}$
Total	$(L \cdot n_s \cdot k_s) - 1$		

between session estimates based on the mean values within session, which are estimates of $\sigma^2_{\text{Between sessions}} + \frac{\sigma^2_{\text{Within session}}}{k_s}$, were shown not to vary significantly (see Section 8.3.5.2), we can reasonably conclude that the total ($\sigma^2_{\text{Within session}} + \sigma^2_{\text{Between sessions}}$) can be pooled across the levels.

Obtaining a valid confidence interval for the IP, based on estimates of the variance components pooled across levels, requires careful consideration of the (possible) interaction between TRP level and session. That is, the analysis should determine if there is variability in the data due to the interaction between session and level. If all replicates are truly independent, then it is likely that the interaction variance will be negligible. However, if the interaction variance is large, then this is likely to be the result of a study design characteristic that is creating level specific effects within a session. Therefore, when the interaction variance is large, a root cause analysis may be beneficial; a plot of the original data is likely to inform this analysis.

Table 8.5 gives the full analysis of variance table including the interaction term. We refer to [8] for the formulae for the expected values of the mean sums of squares. The estimate of the interaction variance is given by:

$$\hat{\sigma}^2_{\text{Interaction}} = \frac{(\text{MSS}_{\text{Interaction}} - \text{MSS}_{\text{Within}})}{k_s}.$$

If $\hat{\sigma}^2_{\text{Interaction}}$ is found to be small, then it can be ignored.

Table 8.6 shows the simplified analysis of variance table, where the interaction variance is assumed to be zero. In this case, the variance components are estimated as:

- Within session variance:

$$\hat{\sigma}^2_{\text{Within session}} = \text{MSS}_{\text{Within}};$$

TABLE 8.6

Variance components calculations for $\widehat{\text{IP}}$, where an interaction is assumed not to exist. DF = degrees of freedom, MSS = mean sum of squares and $\mathbb{E}[\text{MSS}]$ = expected value of mean sum of squares.

Source	DF	MSS	$\mathbb{E}[\text{MSS}]$
Between levels	$L - 1$	$\text{MSS}_{\text{Levels}}$	Not relevant – fixed effect
Between sessions	$n_s - 1$	$\text{MSS}_{\text{Between}}$	$L \cdot k_s \cdot \sigma^2_{\text{Between session}} + \sigma^2_{\text{Within session}}$
Within session	$(L \cdot n_s \cdot k_s) - L - n_s + 1$	$\text{MSS}_{\text{Within}}$	$\sigma^2_{\text{Within session}}$
Total	$(L \cdot n_s \cdot k_s) - 1$		

- Between session variance:

$$\hat{\sigma}^2_{\text{Between sessions}} = \frac{(\text{MSS}_{\text{Between}} - \text{MSS}_{\text{Within}})}{L \cdot k_s};$$

- Total variance:

$$\hat{\sigma}^2_{\text{Between sessions}} + \hat{\sigma}^2_{\text{Within session}} = \text{MSS}_{\text{Between}} \frac{1}{L \cdot k_s} + \text{MSS}_{\text{Within}} \left(1 - \frac{1}{L \cdot k_s}\right).$$

The pooled estimate of intermediate precision is given by:

$$\widehat{\text{IP}} = 100 \times \left\{ \text{antilog} \left[\sqrt{\text{MSS}_{\text{Between}} \frac{1}{L \cdot k_s} + \text{MSS}_{\text{Within}} \left(1 - \frac{1}{L \cdot k_s}\right)} \right] - 1 \right\} \%.$$

The upper confidence limit for IP can be found by following Section 3.4.2.

One method run per session

When $k_s = 1$, the pooled estimate of the total variance in this case is:

$$\hat{\sigma}_{\text{Between sessions}} + \hat{\sigma}_{\text{Within session}} = s^2$$

$$= \frac{\sum_{l=1}^{L} \sum_{i=1}^{n_s} \left(\widehat{\log \text{RP}}_{i \cdot l} - \widehat{\log \text{RP}}_{\cdot \cdot l} \right)^2}{L \cdot (n_s - 1)}.$$

The upper one-sided 95% confidence limit for IP is:

$$100 \times \left\{ \text{antilog} \left[s \sqrt{\frac{L(n_s - 1)}{\chi^2_{0.05, L \cdot (n_s - 1)}}} \right] - 1 \right\} \%.$$

8.3.6 Calculation of tolerance intervals using validation study data

The calculation of tolerance intervals using validation study data can be conducted using the formulae provided in Sections 8.2.2.2 and 8.2.2.3 [7]. We reproduce the formula for the tolerance interval here without the manufacturing distribution:

$$\log\left(\mathrm{TRP}\right) + \log\left(\frac{\widehat{RB}}{100} + 1\right) \pm$$

$$z_{0.995} \times \sqrt{\frac{\left[\left(\hat{\sigma}^2_{\text{Between sessions}} + \hat{\sigma}^2_{\text{Within session}}\right)/n_F\right]\left(1 + \frac{1}{n}\right) \times df}{\chi^2_{0.05;\ df}}}. \qquad (8.8)$$

To incorporate the manufacturing distribution, a factor of $(1 - P)$ should be included in the denominator.

8.3.6.1   Tolerance interval for a given TRP: variance estimated per TRP level

When the variances are estimated for each individual level (i.e., have not been pooled across the TRP levels) the tolerance interval for the reportable value is given by Equation (8.8) replacing the sample size, n, with:

$$n = n_s \cdot k_s,$$

and df by the Satterthwaite degrees of freedom:

$$df = \frac{\left(\hat{\sigma}^2_{\text{Between sessions}} + \hat{\sigma}^2_{\text{Within session}}\right)^2 \times k_s^2}{\mathrm{MSS}_{\text{Between}}^2/(n_s - 1) + (k_s - 1)\,\mathrm{MSS}_{\text{Within}}^2/n_s}.$$

When $k_s = 1$ this reduces to:

$$df = n_s - 1.$$

8.3.6.2   Tolerance interval for a given TRP: variance estimated pooling across TRP levels

When the variance estimates have been pooled across the TRP levels, the tolerance interval is given by Equation (8.8) replacing the sample size n with:

$$n = n_s \cdot k_s,$$

and degrees of freedom:

$$df = \frac{\left(\hat{\sigma}^2_{\text{Between sessions}} + \hat{\sigma}^2_{\text{Within session}}\right)^2 \times (L \cdot k_s)^2}{\mathrm{MSS}_{\text{Between}}^2/(n_s - 1) + \left[(L \cdot k_s) - 1\right]^2 \mathrm{MSS}_{\text{Within}}^2/[(L \cdot n_s \cdot k_s) - L - n_s + 1]}.$$

For a tolerance interval for the relative potency of a lot from the manufacturing distribution, this formula for the degrees of freedom also holds, but the sample size is:

$$n = L \cdot n_s \cdot k_s.$$

8.4   Validation example 1

This example is for a bioassay, where only one method run is possible per day for an operator, i.e., per session. Within session variability is not of interest because it cannot be separately evaluated for practical reasons and is of no use in format design. The following assumptions are made.

- The true RB is 3% and the true IP is 14%, estimated from qualification studies;

- The specification limits for RP are (0.70, 1.43);

- The manufacturing process:

    - Is centred at RP = 1;

    - Has variance 20% of the total variance of the release assay;

- The requirement for the release procedure is that the proportion of manufactured lots falling OOS is to be no more than 1%.

These assumptions lead to the following for the release procedure (see Section 3.4.3):

- ATP requirements:

    - Relative bias within $(-17\%, 20\%)$;

    - Intermediate precision no more than 20%.

- Release format is 8 sessions, each including one method run;

    - This means that the proportion of OOS results for the manufacturing process is less than 1% even if the RB and IP are at the limits of the ATP requirements.

8.4.1 Study design

Five TRP levels will be tested:

$$\{0.50,\ 0.71,\ 1.00,\ 1.41,\ 2.00\}.$$

We will assume that a method run is capable of evaluating five test samples at once. Each validation study session will consist of one method run on one day by one operator. We assume that, in practice, the study sessions will be designed in such a way that the ruggedness factors are reflected. To ensure that the entire extent of likely variability is represented an experimental design can be used. We set the bioassay acceptance limits in line with the ATP as:

- Accuracy: Two-sided 90% confidence interval for RB lies within $(-17\%, 20\%)$;

- Precision: One-sided upper 95% confidence limit for IP lies below 20%.

    - The criterion for pooling the IP estimates across the levels will be that Bartlett's test for comparing the standard deviations of the mean values per session across the levels is not significant at the 5% level. If this is met then a pooled estimate of IP will be calculated as well as individual IP values for each level.

- Linearity: 90% confidence interval for the slope of the linear model for lies within (0.8, 1.25).

    - If the linearity criterion is met, then a pooled estimate of RB will be calculated as well as individual RB estimates for each level.

TABLE 8.7

Simulated results for RP for an example validation study. For this example, there are 10 measurements of RP for each of five TRPs.

Study session	Level: TRP				
	1: 0.50	2: 0.71	3: 1.00	4: 1.41	5: 2.00
1	0.540	0.707	1.037	2.057	2.171
2	0.503	0.595	1.009	1.875	2.147
3	0.494	0.711	1.463	1.510	1.577
4	0.502	0.856	1.020	1.680	1.758
5	0.535	0.773	1.044	1.561	1.876
6	0.570	0.863	1.058	1.367	2.001
7	0.449	0.728	0.781	1.404	2.345
8	0.401	0.554	1.088	1.329	2.392
9	0.491	0.714	1.075	1.617	1.943
10	0.536	0.658	1.157	1.431	2.164

Tables 8.2 and 8.3 both show that we need 10 study runs per target level to achieve a 90% chance of passing the accuracy criterion and a 90% chance of passing the precision criterion pooled across levels.

An alternative to the criteria provided above would be to set a validation acceptance criterion for the tolerance interval for reportable value. If the $(0.99, 0.95)$ tolerance interval falls within the specification limits then we estimate that at least 99% of reportable values will fall in specification with 95% confidence. As this is equivalent to no more than 1% of reportable values falling OOS, a reasonable criterion for the $(0.99, 0.95)$ tolerance interval for the reportable value (for the manufacturing distribution) is that it falls in specification. It would also be possible to place the same criterion on the $(0.99, 0.95)$ tolerance interval for the reportable value for lots with a true potency of 1.

The $(0.99, 0.95)$ tolerance intervals for lots with true potency values far outside the specification limits should fall entirely outside the specification limits. In this case, we can state with 95% confidence that fewer than 1% of such lots will have a reportable value that falls in specification, that is, will wrongly pass the release specification.

8.4.2 Results and analysis

Table 8.7 shows the validation study design, with the results included. The results were simulated as follows for a 3% true RB and a 14% true IP. At each TRP level, 10 random values from a normal distribution with mean value $\log(\text{TRP}) + \log(1.03)$ and standard deviation $\log(1.14)$ were generated. These were then back-transformed and rounded to three decimal places to provide the simulated values of RP; these rounded RP values were used for all calculations. Figure 8.2 shows the relative potency values for each run of the method plotted against their target relative potency values, both on the log scale.

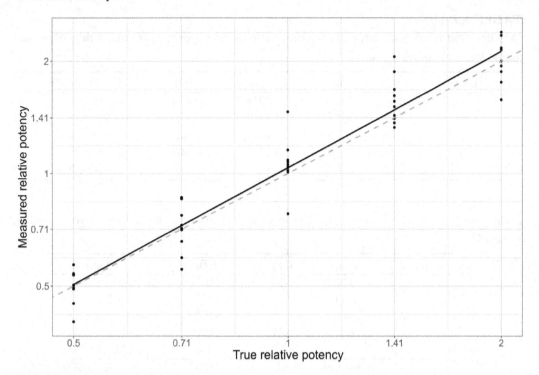

FIGURE 8.2

Relative potency values for 10 sessions with one method run each at each of 5 true potency levels plotted against their true relative potency on the log scale. The line of best fit is shown as a solid line in black and the line of equality is shown as a dashed grey line. The variance appears consistent across the true potency levels.

Note the following:

- $L = 5$;

- $n_s = 10$;

- $k_s = 1$.

8.4.2.1 Accuracy and linearity

Table 8.8 shows the estimate of the RB, along with its 90% confidence interval, estimated using the level-specific standard deviation.

In order to test if the standard deviations can be pooled across the five TRP levels, the pooled standard deviation is calculated as:

$$s = \sqrt{\frac{\sum_{l=1}^{L} \sum_{i=1}^{n_s} \left(\widehat{\log \mathrm{RP}}_{i \cdot l} - \widehat{\log \mathrm{RP}}_{\cdot \cdot l} \right)^2}{L \cdot (n_s - 1)}}$$

$$= 0.1346.$$

TABLE 8.8

Estimate and confidence intervals for the accuracy based on the standard deviation estimated for each individual level.

Level	TRP	$\widehat{\text{RP}}_l$	s_l	$\widehat{\text{RB}}_l$	90% CI for RB_l
1	0.50	0.50	0.1018	0.0%	$(-5.8\%, 6.0\%)$
2	0.71	0.71	0.1412	-0.1%	$(-7.9\%, 8.5\%)$
3	1.00	1.06	0.1528	6.2%	$(-2.8\%, 16.0\%)$
4	1.41	1.57	0.1412	11.2%	$(2.5\%, 20.7\%)$
5	2.00	2.02	0.1300	1.1%	$(-6.2\%, 9.0\%)$

TABLE 8.9

Estimate and confidence intervals for the accuracy based on the pooled standard deviation, $s = 0.1346$.

Level	TRP	$\widehat{\text{RP}}_l$	$\widehat{\text{RB}}_l$	90% CI for RB_l
1	0.50	0.50	0.0%	$(-6.9\%, 7.4\%)$
2	0.71	0.71	-0.1%	$(-7.0\%, 7.3\%)$
3	1.00	1.06	6.2%	$(-1.1\%, 14.1\%)$
4	1.41	1.57	11.2%	$(3.6\%, 19.5\%)$
5	2.00	2.02	1.1%	$(-5.9\%, 8.6\%)$
Pooled across levels			3.6%	$(0.4\%, 7.0\%)$

The Bartlett's test statistic is given by:

$$\chi^2 = \frac{(n_s - 1)^2 \left[L \cdot \log(s^2) - \sum_{l=1}^{L} \log(s_l^2) \right]}{(n_s - 1) + (L+1)/3L}$$

$$= 1.531.$$

The critical value of the χ^2 distribution with 4 degrees of freedom for a 5% level test is $\chi^2_{0.95,4} = 9.488$. The test statistic, 1.531 is less than 9.488, and so there is no evidence of a significant difference amongst the standard deviations across the five levels. Therefore, they may be pooled.

Table 8.9 shows the confidence intervals for RB at each level based on the pooled standard deviation of 0.1346. In general, the confidence intervals based on this pooled estimate tend to be narrower than when the standard deviation is estimated for each level separately (Table 8.8). This is because all data values have been used to estimate the variability.

Table 8.10 shows the estimates of intercept and slope for the linear model relating log RP to log TRP. The 90% confidence interval for the slope is $(0.971, 1.103)$ which falls entirely within the stated limits of $(0.80, 1.25)$. This supports the calculation of a single estimate of relative bias, pooled across the levels.

The pooled estimate of RB can be found using the methods provided in Section 8.3.5.3 and is 3.6% with a 90% confidence interval of $(0.4\%, 7.0\%)$.

TABLE 8.10

Estimate and confidence interval for the parameters of the linearity assessment.

Parameter	Estimate	90% CI
Slope	1.04	(0.971, 1.103)
Intercept	0.04	(0.003, 0.068)

TABLE 8.11

Estimate and upper confidence limit for the precision.

		Standard deviation, s_l		Intermediate precision, IP_l	
Level	TRP	Estimate	Upper 95% CL	Estimate	Upper 95% CL
1	0.50	0.1018	0.1675	10.7%	18.2%
2	0.71	0.1412	0.2324	15.2%	26.2%
3	1.00	0.1528	0.2515	16.5%	28.6%
4	1.41	0.1412	0.2324	15.2%	26.2%
5	2.00	0.1300	0.2139	13.9%	23.8%
Pooled across levels		0.1346	0.1631	14.4%	17.7%

TABLE 8.12

Summary of the (0.99, 0.95) tolerance intervals for each level separately and for the manufacturing processing assuming it is centred at 1 with variance accounting for 20% of the release assay variance.

Level	Given TRP	(0.99, 0.95) tolerance intervals
1	0.50	(0.43, 0.58)
2	0.71	(0.61, 0.83)
3	1.00	(0.91, 1.24)
4	1.41	(1.34, 1.83)
5	2.00	(1.71, 2.36)
Manufacturing process		(0.86, 1.20)

8.4.2.2 Intermediate precision

In the case where $k_s = 1$, the estimates of IP can be calculated directly from the within-level standard deviations, s_l, as shown in Table 8.8. Their confidence limits are calculated as shown in Equations (8.6) and (8.7). Table 8.11 shows the estimate of the standard deviation and IP for each potency level along with their one-sided upper 95% confidence limits. The confidence limit for the estimate pooled across levels was calculated using the chi-squared distribution and the Satterthwaite degrees of freedom.

8.4.2.3 Tolerance intervals

Table 8.12 and Figure 8.3 show the (0.99, 0.95) tolerance intervals for the results of the bioassay for tests of samples conditional on their true relative potency values, and for all results from the manufacturing process assuming that the process is centred at RP = 1 and has variance which accounts for 20% of the total variance of the reportable values, based

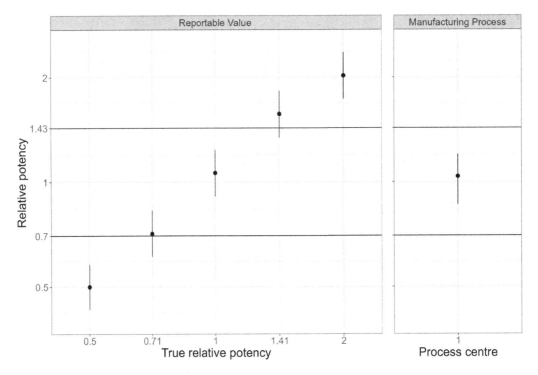

FIGURE 8.3

(0.99, 0.95) tolerance intervals for lots of given true potency, and for the manufacturing process.

on a release format of 8 sessions, each including one method run. The tolerance intervals were calculated using the pooled variance estimate throughout.

8.4.3 Conclusions

Table 8.13 summarises the results of the first validation study example. We conclude the following:

1. Linearity of the assay is demonstrated: the 90% confidence interval for the slope of the regression falls within (0.80, 1.25).

2. The relative bias is acceptable based on the pooled estimate of RB: the 90% confidence interval falls between (−17%, 20%).

3. The intermediate precision is acceptable based on the pooled estimate of IP: the 95% upper confidence limit falls below 20%.

Therefore, the assay has been validated for samples with potency in the range of (0.5, 2.0).

If instead the assay validation was based on tolerance intervals for the reportable value we would conclude the following with 95% confidence:

1. At least 99% of reportable values for the manufacturing process will fall within the specification limits of (0.70, 1.43).

TABLE 8.13

Summary of results for example validation study. The 90% confidence intervals for the RB at each potency level are based on the pooled estimate of the standard deviation, $s = 0.1346$.

Level	TRP	RB	90% CI for RB	IP	95% UCL for IP	(0.99, 0.95) TI
1	0.50	0.0%	(−6.9%, 7.4%)	10.7%	18.2%	(0.43, 0.58)
2	0.71	−0.1%	(−7.0%, 7.3%)	15.2%	26.2%	(0.31, 0.83)
3	1.00	6.2%	(−1.1%, 14.1%)	16.5%	28.6%	(0.91, 1.24)
4	1.41	11.2%	(3.6%, 19.5%)	15.2%	26.2%	(1.34, 1.83)
5	2.00	1.1%	(−5.9%, 8.6%)	13.9%	23.8%	(1.73, 2.36)
Pooled across levels		3.6%	(0.4%, 7.0%)	14.4%	17.7%	(0.86, 1.20)

2. At least 99% of reportable values for lots with true potency of 1 will fall within the specification limits of (0.70, 1.43).

3. Fewer than 1% of reportable values for lots with true potency of 0.5 or 2.0 will fall within the specification limits of (0.70, 1.43).

8.5 Validation example 2

This example is for a bioassay, where more than one method run is possible per day for an operator. For the bioassay to be validated, we make the following assumptions.

- The true RB is 3% and the true IP is 8%, estimated from qualification studies;

- The specification limits for RP are (0.70, 1.43);

- The manufacturing process:

  - Is centred at RP = 1;

  - Has variance 20% of the total variance of the release assay;

- The requirement for the release procedure is that the proportion of manufactured lots falling OOS is to be no more than 1%;

- The requirement for the release procedure is that the proportion of manufactured lots falling OOS lots is to be no more than 1%.

These assumptions lead to the following for the release procedure:

- ATP requirements:

  - Relative bias within (−17%, 20%);

  - Intermediate precision no more than 12%;

- Format is 3 sessions, each including 1 method run:

  - This means that the proportion of OOS results for the manufacturing process is less than 1% even if the RB and IP are at the limits of the ATP requirements.

8.5.1   Study design

Again, we will assume that a method run is capable of evaluating at least five test samples at once. Each validation study run will consist of 3 method runs on one day by one operator. We set the bioassay acceptance limits in line with the ATP as:

- Accuracy: two-sided 90% confidence interval for RB lies within $(-17\%, 20\%)$;

- Precision: one-sided upper 95% confidence limit for IP lies below 12%.

  - The criterion for pooling the IP estimates across the levels will be that: the Bartlett's test for comparing the SDs of the mean values per session across the levels and the Bartlett's test for comparing the within session SDs across the session/level combinations are both non-significant at the 5% level. If this is met then a pooled estimate of IP will be calculated as well as individual IP values for each level.

- Linearity: 90% confidence interval for the slope of the linear model fitted to the within session means of the log RP for lies within $(0.8, 1.25)$.

  - If the linearity criterion is met, then a pooled estimate of RB will be calculated as well as individual RB estimates for each level.

Table 8.2 shows that we need a sample size of at least 6 study sessions to achieve a 90% chance of passing the accuracy criterion at any level individually. Table 8.3 shows that we need at least 7 study runs per target level to achieve a 90% chance and of passing the precision criterion pooled across levels. Taking the larger of these two the study needs to include at least 7 study sessions.

8.5.2   Results and analysis

Table 8.14 shows the validation study design, with the results included. The results for a 3% true RB and an 8% true IP were simulated as follows. At each of five TRP levels, three measurements per session, for a total of seven sessions were simulated. The mean value for each simulated measurement was calculated as $\log(\text{TRP}) + \log(1.03)$ to reflect a RB of 3%. The between session error term was simulated from a normal distribution with mean 0 and variance 0.004; this error term was shared by all measurements within a session. Next, a random error term for each individual measurement was simulated from a normal distribution with mean 0 and variance 0.002. Together, these errors have a total variance of 0.006 resulting in an intermediate precision of 8%. Each measurement was found as the sum of the mean and two error terms. Finally, the measurements were back transformed to

TABLE 8.14

Simulated results for RP for an example validation study. For this example, there are 3 measurements of RP for each of 7 sessions and each of five TRPs.

Session	Session run	1: 0.50	2: 0.71	3: 1.00	4: 1.41	5: 2.00
				Level: TRP		
	1	0.497	0.721	0.984	1.318	1.999
1	2	0.509	0.700	0.958	1.414	2.123
	3	0.494	0.719	0.989	1.482	2.108
	Geomean	0.500	0.713	0.977	1.403	2.076
	1	0.542	0.828	1.147	1.686	2.383
2	2	0.549	0.789	1.055	1.693	2.277
	3	0.571	0.873	1.209	1.655	2.423
	Geomean	0.554	0.829	1.135	1.678	2.360
	1	0.534	0.777	1.014	1.484	2.216
3	2	0.562	0.811	1.086	1.718	2.083
	3	0.595	0.766	1.082	1.557	2.006
	Geomean	0.563	0.784	1.060	1.583	2.100
	1	0.525	0.642	0.903	1.222	1.983
4	2	0.468	0.700	0.993	1.424	1.783
	3	0.466	0.688	0.906	1.323	1.852
	Geomean	0.486	0.676	0.933	1.320	1.871
	1	0.533	0.701	1.021	1.500	1.955
5	2	0.525	0.759	1.068	1.454	1.987
	3	0.505	0.723	0.980	1.488	2.023
	Geomean	0.521	0.727	1.022	1.481	1.988
	1	0.490	0.774	1.121	1.435	2.022
6	2	0.525	0.732	1.052	1.430	1.946
	3	0.500	0.712	0.948	1.360	2.041
	Geomean	0.505	0.739	1.038	1.408	2.003
	1	0.560	0.677	1.054	1.433	2.155
7	2	0.511	0.777	1.013	1.502	2.109
	3	0.508	0.710	1.034	1.501	2.032
	Geomean	0.526	0.720	1.034	1.478	2.098

provide the simulated values of RP; these were rounded to three decimal places. Note the following:

- $L = 5$;

- $n_s = 7$;

- $k_s = 3$.

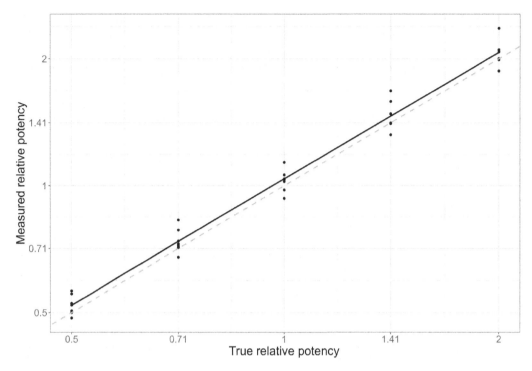

FIGURE 8.4

Relative potency values for 7 sessions, as the geometric mean of three method runs, at each of 5 levels plotted against their true relative potency, on the log scale. The line of best fit is shown as a solid black line and the line of equality is shown as a dashed grey line.

TABLE 8.15

Estimate and confidence intervals for the accuracy based on the standard deviation estimated for each individual level.

Level	TRP	$\widehat{RP}_l$	s_l	$\widehat{RB}_l$	90% CI for RB_l
1	0.50	0.52	0.0540	4.3%	(0.2%, 8.5%)
2	0.71	0.74	0.0670	4.2%	(−0.8%, 9.5%)
3	1.00	1.03	0.0618	2.7%	(−1.9%, 7.4%)
4	1.41	1.47	0.0802	4.6%	(−1.4%, 10.9%)
5	2.00	2.07	0.0713	3.3%	(−2.0%, 8.9%)

Figure 8.4 shows the relative potency values for each run of the method plotted against their target relative potency values, both on the log scale.

8.5.2.1 Accuracy and linearity

Table 8.15 shows the relative bias at each level along with its 90% CI based on the level-specific variance estimates.

TABLE 8.16

Estimate and confidence intervals for the accuracy based on the pooled standard deviation, $s = 0.0674$.

Level	TRP	$\widehat{RP}_l$	$\widehat{RB}_l$	90% CI for RB_l
1	0.50	0.52	4.3%	$(-0.1\%, 8.9\%)$
2	0.71	0.74	4.2%	$(-0.2\%, 8.8\%)$
3	1.00	1.03	2.7%	$(-1.7\%, 7.2\%)$
4	1.41	1.47	4.6%	$(0.2\%, 9.2\%)$
5	2.00	2.07	3.3%	$(-1.1\%, 7.9\%)$
	Pooled across levels		3.8%	$(1.8\%, 5.9\%)$

TABLE 8.17

Estimate and confidence interval for the parameters of the linearity assess- ment.

Parameter	Estimate	90% CI
Slope	1.00	$(0.958, 1.034)$
Intercept	0.04	$(0.019, 0.056)$

In order to test if the standard deviations can be pooled across the five TRP levels, the pooled standard deviation is calculated as:

$$s = \sqrt{\frac{\sum_{l=1}^{L} \sum_{i=1}^{n_s} \left(\overline{\log RP}_{i \cdot l} - \overline{\log RP}_{\cdot \cdot l}\right)^2}{L \cdot (n_s - 1)}}$$

$$= 0.0674.$$

The Bartlett's test statistic:

$$\chi^2 = \frac{(n_s - 1)^2 \left[L \cdot \log(s^2) - \sum_{l=1}^{L} \log(s_l^2)\right]}{(n_s - 1) + (L+1)/3L}$$

$$= 0.9775.$$

The critical value of the χ^2 distribution with 4 degrees of freedom for a 5% level test is 9.488. Therefore, the standard deviations do not differ significantly and they may be pooled across the five levels. Table 8.16 shows the confidence intervals for RB at each level based on the pooled s value of 0.0674.

Table 8.17 shows the estimates of intercept and slope for the linear model relating log RP to log TRP. The 90% confidence interval for the slope is $(0.958, 1.034)$ which falls entirely within the stated limits of $(0.80, 1.25)$. This supports the calculation of a single estimate of RB, pooled across the levels. The pooled estimate of RB is 3.8% with a 90% confidence interval of $(1.8\%, 5.9\%)$.

8.5.2.2 Intermediate precision

In the case where $k_s > 1$, the estimates of IP must be calculated using variance components analysis. Table 8.18 shows the analysis for each level separately.

TABLE 8.18

Estimate and upper confidence limit for the precision for each level analysed separately. The confidence limit for the IP was estimated using the Satterthwaite degrees of freedom. DF = degrees of freedom.

		Level				
		0.50	0.71	1.00	1.41	2.00
Source	DF	Mean Sum of Squares				
Between sessions	6	0.0088	0.0135	0.0115	0.0193	0.0153
Within session	14	0.0019	0.0020	0.0027	0.0024	0.0013
Total	20					
$\hat{\sigma}^2_{\text{Within session}}$		0.0019	0.0020	0.0027	0.0024	0.0013
$\hat{\sigma}^2_{\text{Between sessions}}$		0.0023	0.0038	0.0029	0.0056	0.0046
$\hat{\sigma}^2_{\text{Within session}} + \hat{\sigma}^2_{\text{Between sessions}}$		0.0042	0.0058	0.0056	0.0080	0.0060
$\widehat{\text{IP}}_l$		6.7%	7.9%	7.8%	9.4%	8.0%
Upper one-sided 95% confidence limit for $\widehat{\text{IP}}_l$		10.4%	13.0%	12.1%	15.8%	14.0%

The pooled within session standard deviation across levels is given by:

$$s = \sqrt{\frac{\sum_{l=1}^{L} \sum_{i=1}^{n_s} \sum_{j=1}^{k_s} \left(\overline{\log \text{RP}}_{ijl} - \overline{\log \text{RP}}_{i \cdot l}\right)^2}{L \cdot n_s \cdot (k_s - 1)}}$$

$$= 0.0456.$$

Bartlett's test statistic to compare the within session standard deviations across levels is:

$$\chi^2 = \frac{(k_s - 1)^2 \left[L \cdot n_s \cdot \log(s^2) - \sum_{l=1}^{L} \sum_{i=1}^{n_s} \log(s_{il}^2)\right]}{(k_s - 1) + \left[(L \cdot n_s) + 1\right]/(3L \cdot n_s)}$$

$$= 27.242.$$

The critical value of the χ^2 distribution with 34 degrees of freedom for a 5% level test is $\chi^2_{0.95,34} = 48.602$. Therefore, the within session standard deviations do not differ significantly and they may be pooled across the five levels.

Table 8.19 shows the full variance components analysis including the interaction term. In this case, the interaction variance is estimated to be 0. Therefore, for the estimation of the confidence limit for the pooled IP, the interaction term can be ignored. Table 8.20 provides the simplified variance components analysis without the interaction term. Here, the estimate of the IP is 8.0% with an upper confidence limit of 12.3%.

8.5.2.3   Tolerance intervals

Table 8.21 and Figure 8.5 show the (0.99, 0.95) tolerance intervals for the results of the bioassay for tests of samples conditional on their true relative potency values. Also shown are the results from the manufacturing process assuming that the process is centred at an RP of 1 and has variance which accounts for 20% of the total variance of the reportable values, based on a release format of 3 sessions, each including one method run.

TABLE 8.19

Estimates of the variance components including the interaction term. DF = degrees of freedom, MSS = mean sum of squares.

Source	DF	MSS
Between levels	4	6.22744
Between sessions	6	0.06146
Levels × session interaction	$(5-1) \cdot (7-1) = 24$	0.00169
Within session	$5 \cdot 7 \cdot 3 - 5 - 7 + 1 = 70$	0.00208
Total	$5 \cdot 7 \cdot 3 - 1 = 104$	

$$\hat{\sigma}^2_{\text{Within session}} = 0.00208$$
$$\hat{\sigma}^2_{\text{Interaction}} = 0$$
$$\hat{\sigma}^2_{\text{Between sessions}} = 0.00398$$
$$\hat{\sigma}^2_{\text{Between sessions}} + \hat{\sigma}^2_{\text{Within session}} = 0.00606$$

TABLE 8.20

Estimates of the within and between session variability and intermediate precision with upper confidence limit, pooled across levels. DF = degrees of freedom, MSS = mean sum of squares.

Source	DF	MSS
Between levels	4	6.22744
Between sessions	6	0.06146
Within session	$5 \cdot 7 \cdot 3 - 5 - 7 + 1 = 94$	0.00198
Total	$5 \cdot 7 \cdot 3 - 1 = 104$	

$$\hat{\sigma}^2_{\text{Within session}} = 0.00198$$
$$\hat{\sigma}^2_{\text{Between sessions}} = 0.00397$$
$$\hat{\sigma}^2_{\text{Between sessions}} + \hat{\sigma}^2_{\text{Within session}} = 0.00595$$
$$\widehat{\text{IP}} = 8.0\%$$
Upper one-sided 95% confidence limit for $\widehat{\text{IP}}$: 12.3%

TABLE 8.21

Summary of the (0.99, 0.95) tolerance intervals for each level separately and for the manufacturing processing assuming it is centred at 1 with variance accounting for 20% of the release assay variance.

Level	Given TRP	(0.99, 0.95) tolerance intervals
1	0.50	(0.44, 0.62)
2	0.71	(0.62, 0.88)
3	1.00	(0.86, 1.22)
4	1.41	(1.24, 1.76)
5	2.00	(1.73, 2.46)
Manufacturing process		(0.88, 1.23)

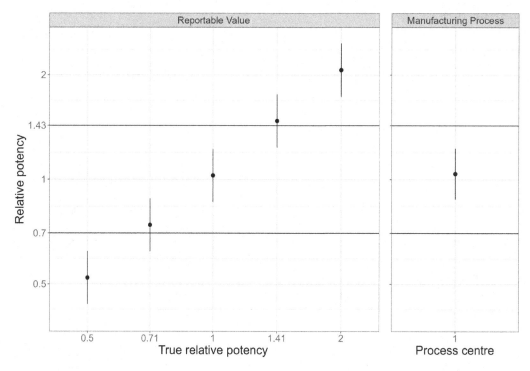

FIGURE 8.5

(0.99, 0.95) tolerance intervals for lots of given true potency, and for the manufacturing process.

TABLE 8.22

Summary of results for example validation study. The 90% confidence interval for RB is based on the pooled estimate of the standard deviation, $s = 0.0674$.

Level	TRP	RB	95% CI for RB	IP	95% UCL for IP	(0.99,0.95) TI
1	0.50	4.3%	(−0.1%, 8.9%)	6.7%	10.4%	(0.44, 0.62)
2	0.71	4.2%	(−0.2%, 8.8%)	7.9%	13.0%	(0.62, 0.88)
3	1.00	2.7%	(−1.7%, 7.2%)	7.8%	12.1%	(0.86, 1.22)
4	1.41	4.6%	(0.2%, 9.2%)	9.4%	15.8%	(1.24, 1.76)
5	2.00	3.3%	(−1.1%, 7.9%)	8.0%	14%	(1.73, 2.46)
Pooled across levels		3.8%	(1.8%, 5.9%)	8.0%	12.3%	(0.88, 1.23)

8.5.3 Conclusions

Table 8.22 summarises the results of the second validation study example.

We can conclude the following:

1. Linearity of the assay is demonstrated: the 90% confidence interval for the slope of the regression falls within (0.80, 1.25).

2. The relative bias is acceptable based on the pooled estimate of RB: the 90% confidence interval falls between (−17%, 20%).

3. The intermediate precision is acceptable based on the pooled estimate of IP: the 95% upper confidence limit falls below 12%.

Therefore, we can say that the validated range of the assay is [0.50, 2.00].

We can also conclude the following with 90% confidence:

4. At least 99% of reportable values for lots with true potency of 1 will fall within the specification limits of (0.70, 1.43).

5. Fewer than 1% of reportable values for lots with true potency of 0.5 or 2.0 will fall within the specification limits of (0.70, 1.43).

6. At least 99% of reportable values for the manufacturing process will fall within the specification limits of (0.70, 1.43).

8.6 Chapter summary

- Chapter 8 describes how to test whether the assay developed as described in Chapters 3-7 will meet the analytical target profile (ATP) developed in Chapter 2. This is often called bioassay qualification, verification or validation.

- The validation study must be based on the developed standard operating procedure and the sample size (number of repeats of the standard procedure) designed such that the results are representative of what can be expected in routine use.

- The statistical analysis of the results of the study is explained in detail, including calculation of the relative bias, linearity, intermediate precision and tolerance interval.

- The use of variance components analysis for the precision performance characteristics is explained. Interpretation of the performance characteristics with respect to the ATP is illustrated. Two fully worked examples are included.

9

Monitoring the performance of a bioassay procedure

9.1   Introduction

Once a bioassay has been validated and put into routine use, it is often a requirement of the regulators to monitor the performance of the assay over time. Statistical process control (SPC) techniques provide the methodologies required for objective routine monitoring of the assay. These methods can be useful for the early identification of any shifts, drifts, or changes to various performance characteristics of the assay.

In this chapter, we introduce the concepts of statistical process control and how these concepts can be applied to the monitoring of the performance of a bioassay. The aim of this chapter is to provide an overview and demonstration of a few simple statistical process control techniques. For a more thorough discussion of the topic, we refer to the following textbooks [41, 45, 52].

Statistical process control has a long history, with applications in manufacturing and production dating back to the 1920's with Walter Shewhart at Bell Telephone. These techniques seek to differentiate between common and special causes of variation and can provide early warning signs of any changes to process so that corrective action can be taken. Common causes of variation are resultant from the natural variability in the process and reflect the noise in the day-to-day operation of the process. This source of variation can be quantified using historical data. A special cause of variation is one that is resultant from a fundamental change to the process that may require corrective action. Most often, SPC will make use of control charts that plot process data collected over time to provide a visual representation of the performance of the process. Some control charts are still called Shewhart charts.

Figure 9.1 shows an example of a control chart for the slope parameter, B, of a 4PL curve (the data for this plot can be found in and found in Table 9.2). The individual points represent the slope parameter of each assay. Horizontal lines have been plotted at the mean; at the mean $\pm$ 2 standard deviations; and at the mean $\pm$ 3 standard deviations. These will later be used to define a set of rules that can be used to determine if the assay is in control.

DOI: 10.1201/9781003449195-9

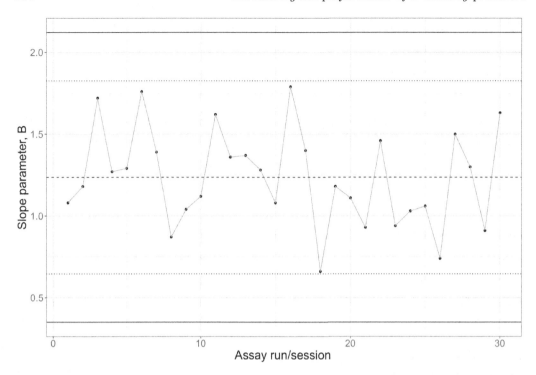

FIGURE 9.1

Control chart for the slope parameter, B of a 4PL curve. The individual points represent the slope parameter of each assay. Horizontal lines have been plotted at the mean (centre, dashed line); at the mean $\pm$ 2 standard deviations (dotted); and the mean $\pm$ 3 standard deviations (solid).

9.2 Monitoring plan development

It is only once the assay is under control and when no further development is being undertaken that a monitoring plan should be implemented. Implementation prior to this stage risks overestimating the variability of the assay as well as introducing potential bias. An overestimation of the variability may result in subsequent missed signals and delays to identifying shifts in assay performance. The introduction of bias may result in the SPC rules being triggered inappropriately.

The development of a monitoring plan begins with choosing the metrics (numeric characteristics of assay performance) to be monitored. The range of metrics available depends on the dose-response relationship, and the most appropriate set will often be context-specific. For example, if the reference standard is known to have a very short shelf life, then monitoring of the C parameter (log EC_{50}) may be critical. However, if the reference standard is known to be very stable then monitoring this parameter may be of lesser importance. There is a balance in ensuring enough metrics are monitored that the performance characteristics of the assay are adequately tracked over time, but that not too many are monitored such that there are many investigations of potential signals.

Following the choice of metrics, the type of control chart to be implemented and any rules for identifying shifts and drifts in the assay performance need to be chosen. We will discuss these choices in subsequent sections. It can be beneficial to collect information on multiple metrics of the assay but only place rules on a few key sensible aspects. Monitoring is a circular process, and the rules can be adjusted if they are not picking up the correct flags.

9.2.1 Metrics

Common metrics for monitoring include assessment of the model fit and parameters of the reference standard curve, as well as the parallelism assessments for the assay control and its relative potency. These components should be stable if the assay is not undergoing any changes. Monitoring of parameters associated with the test sample may not directly provide information on the performance of the assay. The test parameters will also incorporate variability due to the manufacturing process or variability due to the degradation of the product if the assay is used for stability testing. These may result in the false triggering of any SPC rules.

9.2.2 Reference standard curve parameters

The choice of reference standard curve parameters available to be monitored will be determined by the dose-response relationship for the assay. When choosing the parameters to monitor, enough should be chosen to ensure that the assay performance over time can be adequately characterised. However, it is worth noting that many assay metrics will be highly related to one another. Therefore, it may be worth monitoring only a select few. For example, for the 4PL relationship, it would not be necessary to monitor the assay range (difference between the asymptotes), the individual asymptote values themselves and the slope as these parameters will all be highly related to one another.

Straight-line relationship

For assays with a straight-line dose-response relationship, possible metrics include the estimates of the intercept (β_0) and slope (β_1), the root-mean-squared-error and the estimate of a quadratic term.

Tracking the estimate of a quadratic term can indicate if the assay is drifting outside of the linear range. Monitoring of the intercept will flag if, on average, the responses of the reference standard have increased (or decreased), while tracking the slope parameter will indicate if the relationship between dose and response is changing. For example, as the reference standard degrades, it may result in less of a change to the response with increasing dose (i.e., a flattening of the line). Finally, tracking the residual variability as a root-mean-squared-error can help to indicate if the model fit to the data is getting worse or better.

Four-parameter logistic relationship

For assays modelled using a four-parameter logistic curve, possible metrics to monitor include the four parameters of the reference curve fit as well as the root-mean-squared-error. As the parameters of the 4PL curve are highly related, it may be sensible to choose only a subset of the parameters to track, for example, only monitoring the (log) EC_{50} (i.e., either the C parameter or its antilog), the slope parameter, B, and the assay range (asymptote difference).

Monitoring the (log) EC_{50} of the reference standard curve is important to understanding if the potency of the reference standard is changing over time. While this value can be expected to vary from day-to-day according to the variability of the assay, in a stable environment we would not expect to observe systematic changes or drifts. Similar to the straight-line relationship, monitoring of the slope parameter will indicate if the responsiveness of the assay to changes in the dose is drifting over time. Finally, monitoring the assay range may be important as a reduction could indicate that the reference standard is no longer able to elicit the same response on average.

Five-parameter logistic relationship

For an assay with a five-parameter logistic dose-response relationship, monitoring of the parameter E, in addition to the analogous parameters of the four-parameter logistic model can be beneficial. Observing changes to the E parameter over time would indicate a change in the symmetry of the assay dose-response relationship.

Binary response

The binary models presented in this book all assume a straight-line relationship between the logit (or probit) transformed response and the dose. Possible metrics to monitor include the estimates of the intercept (β_0) and slope (β_1), as well as the estimate of a quadratic term for the reference standard fit. Note that assessing the model fit by the root-mean-squared-error is not appropriate for binary data.

9.2.3   Assay control

It can be beneficial to monitor the criteria used for the assessment of parallelism between the reference standard and assay control, as well as the assay control relative potency. For example, if the assay control is made up of different material from the reference standard, any changes to these parameters could be indicative of degradation of the samples. The available metrics associated with the assessment of parallelism will depend on the analysis of the assay. We have discussed many options in Chapter 7. These metrics could, for example, include the any of the comparisons of the individual parameters.

9.2.4   Summary

Table 9.1 provides possible metrics to monitor for each of the dose-response relationships discussed in this book.

TABLE 9.1

Possible metrics for monitoring for three continuous dose-response relationships (straight-line, 4PL, 5PL) and one binary dose-response relationship (straight line).

Parameter	Continuous data			Binary data
	Straight line	4PL	5PL	Straight line
Reference Standard Parameters				
Location (β_0 or C)	✓	✓	✓	✓
Slope parameter (β_1 or B)	✓	✓	✓	✓
Range (asymptote difference, $D - A$)		✓	✓	
Asymmetry (E)			✓	
Quadratic term	✓			✓
Root-mean-squared-error	✓	✓	✓	
Assay Control				
Relative potency	✓	✓	✓	✓
Parallelism: slope parameter	✓	✓	✓	✓
Parallelism: left asymptote		✓	✓	
Parallelism: right asymptote		✓	✓	
Parallelism: asymmetry			✓	

9.3 Control charts

Control charts provide a visual representation of the performance of a process by plotting data recorded over a period of time. Typically these charts will incorporate control limits representing the typical variability about the centre of the process. Therefore, the initial creation of these charts should be based on data for a process that is already stable and in control. The inclusion of extreme, or anomalous data, may result in an overestimate of the variability or inappropriate placement of the centre line.

There are several choices of control charts and the most appropriate type will depend on the nature of the metric being monitored. More specifically, it will depend on whether the metric is continuous or categorical as well as if the data collected are collected as individual values, or if they are grouped in some way. In general, there are two main ways to monitor the metrics: (i) monitoring of the observed values or (ii) monitoring the variability of the observed values. In this book we will only consider the monitoring of continuous metrics. We will also primarily focus on the monitoring of individually observed values and will briefly discuss the handling of grouped data.

9.3.1 Individual values chart

An individual values chart, sometimes called an I-chart, is perhaps the simplest control chart, and, as the name implies, is a control chart for the observed individual values. To provide an example, this could be the slope parameter, B, of the reference standard for every instance of the assay. An example of an individual values chart is shown in Figure 9.1.

If the data are grouped, for example, if three assay plates are prepared simultaneously within a single session by the same analyst, then this structure should be taken into account when establishing the control chart. In this case we can expect the replicate measurements within an assay session to be more similar to each other, than between assay sessions. In this case, control charts could treat the mean values as individuals to ensure the correct variability is used.

When establishing a control chart for the individual values, it is beneficial to have an understanding of the distribution of the data. A histogram of the data can aid in this purpose and can be used to identify if there are any gaps in the data as well as if the distribution is symmetric. To assess if the data are normally distributed, a normal quantile-quantile plot can be produced and a hypothesis test such as the Shapiro Wilk test can be conducted [56].

An assumption of normality is not critical in the application of statistical process control techniques. That is, even if the data are non-normal, a control chart may have an acceptable performance. If the data are very skewed, it is possible to end up with control limits that are not sensible. For example, a negative lower control limit for a process where the individual data points are restricted to only positive values. Further, if the data are non-normal, the probabilities associated with triggering a rule and the run lengths to first signal may be higher (or lower) than expected depending on the distribution of the data.

When the data have natural boundaries, while it is possible to truncate any limits at the boundaries, this is not always advisable. Truncation at a natural boundary would result in a scenario in which it is impossible to observe any results beyond the control limit. Alternatively, it is possible to work on a transformed scale (for example the log scale) and then back-transform to the observed scale. This would be the natural approach when the data have a lower boundary of zero. Finally a control chart can be built based on a non-normal distribution – this process can be difficult and may not be worth the additional effort [67].

9.3.1.1   Control limits

When establishing control limits it is important that the limits are calculated only once the system is in control. Control limits should only be set based on an in-control process of at least 30 points; however provisional limits may be set based on a shorter series (ideally at least 20 points) and kept under review.

Let X be a variable that has normal distribution with mean μ and standard deviation σ. Here, X will represent any metric that we wish to monitor, for example the slope parameter of the reference standard curve. Then, the probability (assuming a normal distribution) that an observation is no more than three standard deviations from the mean is approximately 99.73%, i.e,

$$Prob(\mu - 3 \times \sigma \leq X \leq \mu + 3 \times \sigma) = 0.9973.$$

Therefore, it is very unlikely that a value will fall more than three standard deviations away from the mean due to background variation alone (probability of about 0.27%). When an individual value falls outside these limits, it is likely resultant of a special cause of variation, or an aberrant result, that warrants investigation, rather than by chance alone.

Similarly, for data that have a normal distribution, approximately 95% of observations fall within two standard deviations of the mean i.e,

$$Prob(\mu - 2 \times \sigma \leq X \leq \mu + 2 \times \sigma) = 0.9545.$$

In other words, approximately 5% of observations can be expected to fall more than two standard deviations from the mean simply by random by chance. Therefore, obtaining multiple consecutive observations, or multiple observations within a short sequence, more than two standard deviations beyond the mean would again be indicative of a special cause of variation.

In general, the mean and the standard deviation of the normal distribution are unknown parameters and must be estimated from the data. There are a number of methods for computing the limits for a control chart; we present one here where the limits are based on the sample mean and sample standard deviation. The resulting control chart is sometimes referred to as a Levey-Jennings chart [33].

We assume that the control limits will be calculated using a data set consisting of n measurements of the metric, labelled as $i = 1, \ldots, n$. Let x_i be the ith measurement of the metric. Then, the sample mean can be calculated as:

$$\bar{x} = \frac{1}{n} \sum_{i=1}^{n} x_i.$$

The sample mean is plotted as the centre line (see Figure 9.1). The sample standard deviation is calculated as:

$$s = \sqrt{\frac{1}{n-1} \sum_{i=1}^{n} (x_i - \bar{x})^2}.$$

The upper and lower control limits are sometimes referred to as alarm limits and are calculated as:

$$\text{Lower and upper alarm limits} = (\bar{x} - 3 \times s, \ \bar{x} + 3 \times s),$$

Alert limits can also be calculated as:

$$\text{Lower and upper alert limits} = (\bar{x} - 2 \times s, \ \bar{x} + 2 \times s).$$

Figure 9.1 shows the alarm and alert limits calculated using these equations as horizontal lines. Alternative definitions of the control limits exist based on the moving range; we will present these in the next section.

9.3.2   Moving range chart

Rather than plotting the individual values themselves, a moving range chart looks at the absolute difference between the current observation (i) and previous observation ($i - 1$), denoted as MR_i, and calculated as:

$$MR_i = |x_i - x_{i-1}|.$$

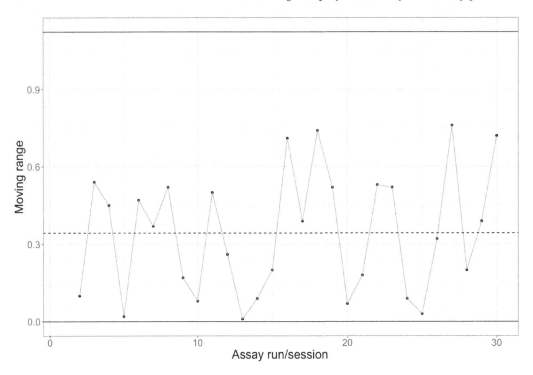

FIGURE 9.2

Control chart for the moving range of the slope parameter of a 4PL curve, B.

Note that when a data set consists of n observations, then there will be $n-1$ moving ranges. Similar to setting a centre-line for the individual values, we can establish a centre line for the moving range by calculating the mean of the differences:

$$\overline{MR} = \frac{\sum_{i=2}^{n} |x_i - x_{i-1}|}{n-1},$$
$$= \frac{\sum_{i=2}^{n} MR_i}{n-1}.$$

Control limits for the moving range can be calculated as follows:

$$\text{Lower alarm limit} = D_2 \times \overline{MR},$$
$$\text{Upper alarm limit} = D_4 \times \overline{MR},$$

where D_2 and D_4 are control limit constants which depend on the number of observations over which the moving range is calculated, here 2 [45]. When the differences are calculated between two consecutive observations as we have defined here, then $D_2 = 0$ and $D_4 = 3.27$. Therefore, the lower control limit will be 0.

Figure 9.2 plots the moving range control chart for the same data presented in Figure 9.1 and found in Table 9.2. No points are outside the alarm limits. We demonstrate the calculation of these limits below.

TABLE 9.2

Example monitoring data set for the slope parameter, B.

Observed value of slope parameter, B										
$n = 1 - 10$	1.08	1.18	1.72	1.27	1.29	1.76	1.39	0.87	1.04	1.12
$n = 11 - 20$	1.62	1.36	1.37	1.28	1.08	1.79	1.40	0.66	1.18	1.11
$n = 21 - 30$	0.93	1.46	0.94	1.03	1.06	0.74	1.50	1.30	0.91	1.63

Moving ranges										
$n = 1 - 9$		0.10	0.54	0.45	0.02	0.47	0.37	0.52	0.17	0.08
$n = 10 - 19$	0.50	0.26	0.01	0.09	0.20	0.71	0.39	0.74	0.52	0.07
$n = 11 - 29$	0.18	0.53	0.52	0.09	0.03	0.32	0.76	0.20	0.39	0.72

TABLE 9.3

Summary statistics and control limits for the individual values chart and moving range chart for the slope data in Table 9.2.

		Count	Mean	Std dev
Observed values of B	$n = 30$		$\bar{x} = 1.2357$	$s = 0.2952$
Moving ranges	$n = 29$		$\overline{MR} = 0.3431$	–

9.3.2.1 Control limits for individual values chart based on moving range

Alternative control limits for the individual values chart can also be calculated based on the average moving range value. The alarm limits are calculated as:

$$\text{Lower and upper alarm limits} = \left(\bar{x} - 3 \times \frac{\overline{MR}}{d_n}, \quad \bar{x} + 3 \times \frac{\overline{MR}}{d_n} \right),$$

and the alert limits as:

$$\text{Lower and upper alert limits} = \left(\bar{x} - 2 \times \frac{\overline{MR}}{d_n}, \quad \bar{x} + 2 \times \frac{\overline{MR}}{d_n} \right),$$

where d_n is a constant based on the number of observations over which the moving range is calculated, in this case $d_2 = 1.128$ [45]. Figure 9.3 shows the control chart for the slope parameter, B, with the alarm limits and alert limits for calculations based on the sample mean and standard deviation and moving range (the data for this plot are in Table 9.2, and we demonstrate the calculations below).

9.3.3 Example: individual values and moving range charts

Table 9.2 contains an example data set consisting of 30 estimates of the slope parameter B for the 4PL curve fit to a reference standard. The differences between each pair of consecutive

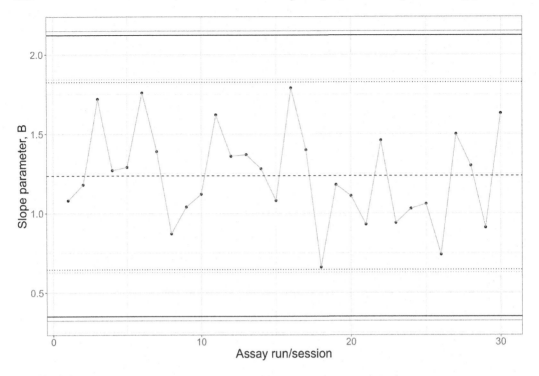

FIGURE 9.3
Control chart for the slope parameter of a 4PL curve, B, with alarm limits (solid) and alert limits (dotted) for calculations based on the sample mean and standard deviation (black) and moving range (grey). The limits based on the moving range are slightly wider for these data.

observations (the moving ranges) have also been recorded. Figures 9.4 and 9.5 show a histogram and normal quantile-quantile plot for these data. Overall, it can be seen that the data are symmetric with no obvious gaps and there is no evidence of non-normality.

Table 9.4 contains the mean and standard deviation of the slope parameter data presented in Table 9.2, the average moving range, as well as a summary of the control limits. When calculated using the sample mean and sample standard deviation, the upper and lower control limits are found as:

$$\text{Lower and upper alarm limits} = (\bar{x} - 3 \times s, \quad \bar{x} + 3 \times s)$$
$$= (1.2357 - 3 \times 0.2952, \quad 1.2357 + 3 \times 0.2952)$$
$$= (0.350, \ 2.121),$$

with alert limits found as:

$$\text{Lower and upper alert limits} = (\bar{x} - 2 \times s, \quad \bar{x} + 2 \times s)$$
$$= (1.2357 - 2 \times 0.2952, \quad 1.2357 + 2 \times 0.2952)$$
$$= (0.645, \ 1.826).$$

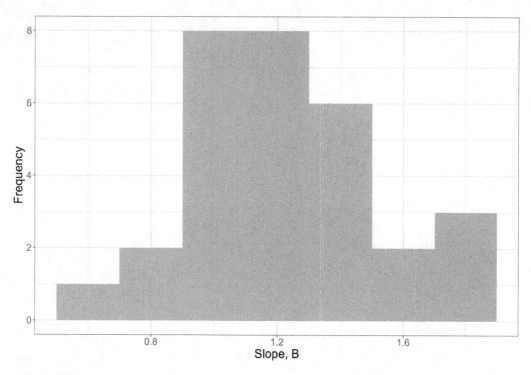

FIGURE 9.4

Histogram of the slope parameter data. There are no obvious gaps in the data and the data are relatively symmetric.

Alternatively, the limits can be calculated using the average moving range as:

$$\text{Lower and upper alarm limits} = \left(\bar{x} - 3\frac{\overline{MR}}{d_n}, \ \ \bar{x} + 3\frac{\overline{MR}}{d_n} \right)$$

$$= \left(1.2357 - 3 \times \frac{0.3431}{1.128}, \ \ 1.2357 + 3 \times \frac{0.3431}{1.128} \right)$$

$$= (0.323, \ 2.148)$$

with alert limits:

$$\text{Lower and upper alert limits} = \left(\bar{x} - 2\frac{\overline{MR}}{d_n}, \ \ \bar{x} + 2\frac{\overline{MR}}{d_n} \right)$$

$$= \left(1.2357 - 2\frac{0.3431}{1.128}, \ \ 1.2357 + 2\frac{0.3431}{1.128} \right)$$

$$= (0.627, \ 1.844).$$

For these data, there is very little difference between the control limits calculated using the standard deviation and the average moving range.

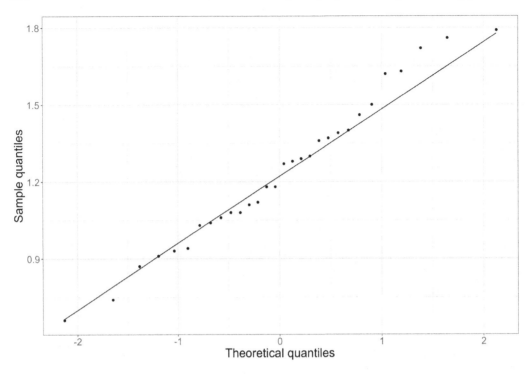

FIGURE 9.5

Normal quantile-quantile plot of the slope parameter data. There is no obvious evidence of non-normality.

TABLE 9.4

Summary statistics and control limits for the individual values chart and moving range chart for the slope data in Table 9.2.

Method	Centre line	Alert limits	Alarm/Control limits
Individual values chart			
Mean and standard deviation	1.2357	(0.350, 2.121)	(0.645, 1.826)
Mean and moving range	1.2357	(0.323, 2.148)	(0.627, 1.844)
Moving range chart			
	0.3431		(0, 1.122)

The control limits for the moving range chart can be found as:

$$\text{Lower and upper alarm limits} = (D_2 \times \overline{MR}, \ D_4 \times \overline{MR})$$
$$= (0 \times 0.3431, \ 3.27 \times 0.3431)$$
$$= (0, \ 1.122).$$

9.3.4 Incorporating subgroup structure in process control

The control charts that we have presented thus far were for individual values. However, in some cases the data will be collected in natural subgroups. For example, consider an assay session where three plates are prepared by the same analyst. Then, the plates prepared within the single session will form a natural subgroup. We present methods to handle this type of data below.

X-bar chart

An x-bar chart is very similar in nature to the individual values chart as previously described. Here a control chart for the sample mean of the subgroup is created rather than the individual values. However, the estimate of variability does not account for the subgroup-to-subgroup variability; it is calculated from the within subgroup variability. Therefore, its application to bioassay is limited. An alternative strategy could be to calculate the control limits using a variance components approach.

Range chart

For data with natural subgroups, the range chart can plot the range (maximum - minimum) for each subgroup.

S-chart

The s-chart monitors the variability inside of each subgroup, calculated as the subgroup standard deviation. As standard deviation calculations can be highly variable when the sample size is small, this type of control chart is recommended when the sample size for each subgroup is at least 10. This type of control chart is also useful when the sample size for each subgroup may differ.

9.3.5 Other charts

There are many other control chart options, for example the cumulative sum (CUSUM) or exponentially weighted moving average (EWMA) charts. For the ith observation, the CUSUM chart calculates the sum of differences from the centre line up to and including the ith observation. A CUSUM chart will result in a signal for a shift sooner than an individual values or x-bar chart. The EWMA chart also incorporates historic information by calculating a weighted average, where more recent observations carry more weight than earlier ones.

9.4 Establishing if the process is under control

Before establishing a control chart, it is important first to understand if the process is in control. If a state of control has not yet been reached, then it is quite likely that any control limits set will be too wide and will fail to miss future signals. There are a number of rules

TABLE 9.5

Rules for establishing if a process is in control as proposed in [45].

1	No values outside alarm limits.
2	No more than 1 in 40 values between the alert and alarm limits.
3	No 2 consecutive values outside the same alert limit.
4	No runs or trends with length of 5 or that also infringe an alert (or alarm) limit.
5	No runs of length 6 or more entirely above or entirely below the mean (centre line).
6	No rising or falling trends of length 6 or more.

that can be used to understand if the process is under control. Table 9.5 contains a set of rules for establishing if a process is in control as proposed in [45].

The first three rules in Table 9.5 are related to the control limits (see Section 9.3.1.1 for discussion on the rationale of these limits). Rules 4–6 refer to concepts of a run and a trend. A run is defined as a series of consecutive points sitting either entirely above the centre line or sitting entirely below the centre line. This pattern would suggest that the mean of the process has shifted upwards (or downwards). A trend is defined as a series of consecutive points where each successive point is larger than the previous for an increasing trend, or smaller than the previous for a decreasing trend. This pattern would indicate a gradual drift upwards or downwards rather than a step-change.

The fifth rule involves looking at the number of consecutive points that sit entirely above or entirely below the centre. Ideally, if the process is under control then we would expect to see the individual observations scattered back and forth about the centre line. However, if the process has shifted, then we may expect many consecutive points all sitting on the same side. The sixth rule is triggered when there is a long increasing or decreasing trend. The fourth rule is triggered when a shorter run or trend also infringes an alert limit.

If any of these rules do not hold, then the process may not be in control. In the case where specific causes can be attributed to extreme points and may be causing the process to be out of control, then the points may be excluded, limits recalculated and the checks for control may be re-run. Applying this set of rules to the data for the slope parameter, B, analysed in Section 9.3.3, we find that the process is in control.

Figures 9.6 shows another data set (not provided) also for a slope parameter. For these data, there is one point outside the alarm limits and two consecutive points that are outside the same alert limit. These two points are also between the alert and alarm limits and represent more than 1 in 40 of the total. According to the rules in Table 9.5 this process is not in control.

9.5 The monitoring process

There are many different sets of rules that can be used in the monitoring of a process. These include the Western Electric rules [15], Westgard rules [66] and the Nelson rules [44]. We summarise the Nelson rules in Table 9.6.

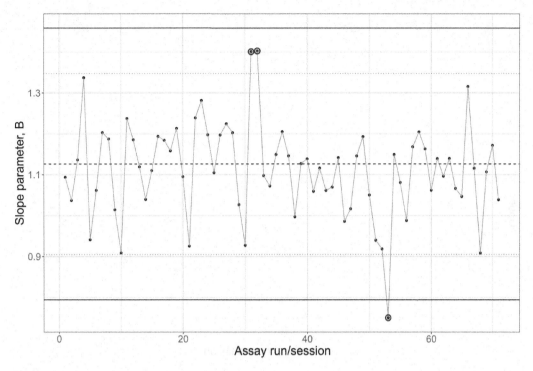

FIGURE 9.6

Example data for the slope parameter of a 4PL curve. There is one point outside the alarm limits and two consecutive points that are outside the same alert limit. These two points are also between the alert and alarm limits and represent more than 1 in 40 of the total.

TABLE 9.6

Nelson rules for monitoring a process [44].

1	One value outside the alarm limits.
2	Nine or more sequential values entirely above or entirely below the centre line.
3	Six consecutively rising or falling values.
4	Fourteen or more consecutive values in an alternating pattern (increase followed by decrease).
5	Two out of three values in a row more beyond the same alert limit.
6	Four out of five sequential values more than one standard deviation from the mean.
7	Fifteen sequential values are no more than 1 standard deviation from the mean.
8	Eight sequential values are all more than 1 standard deviation from the mean, and fall on both sides of the mean.

It is assumed that monitoring will be in place long enough that a failure will be observed. Therefore, it can often be helpful to think about the average run length, or number of observations before the first failure (assuming that the process is truly in control). For example, we expect (approximately) 99.73% of observations to fall within 3 standard deviations of the mean for normally distributed data. Therefore, we would expect about 1 out of every 370 observations to fall outside the alarm limits just by chance.

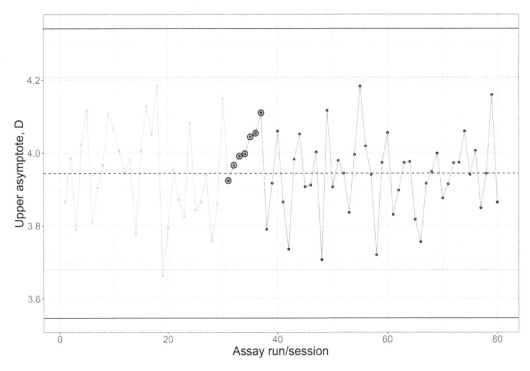

FIGURE 9.7

Example monitoring data for the upper asymptote of a 4PL curve. The control limits were set using the first 30 points (in grey); the remaining data represent ongoing monitoring. These data display an increasing trend of 7 points.

Figures 9.7 and 9.8 illustrate the monitoring process. In both figures, the control limits were set using the first 30 points (shown in grey). The subsequent points represent the ongoing data being monitored. Figure 9.7 has no points outside the alarm limits. However, it displays a trend of 7 points from values 31 to 40; this is a violation of the third Nelson rule. In practice an investigation should be implemented after the sixth value was observed. In Figure 9.8, there is a run of 9 consecutive points entirely below the centre line. This violates the second Nelson rule.

Monitoring too many metrics and/or using too many rules can lead to a high number of investigations, depending on the independence of the metrics. The probabilities of failure will be cumulative, so failures can be frequent even if the likelihood of any one rule being failed is very small. For example, for 5 (independent) metrics each with a 0.27% of chance failure, we would expect to have a false signal for a process truly in control on average every 74 points.

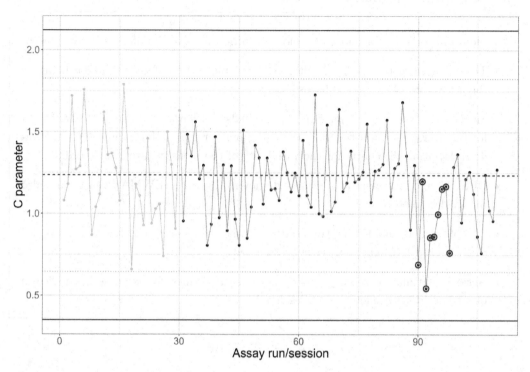

FIGURE 9.8

Example monitoring data for the log EC_{50}, C, parameter of the 4PL curve. The control limits were set using the first 30 points (in grey); the remaining data represent ongoing monitoring. These data display a run of 9 consecutive points entirely below the centre line.

9.6   Chapter summary

- Monitoring the performance of a bioassay procedure over its lifetime is introduced.

- The choice of metrics to be monitored for the bioassay's reference standard and for the assay control sample is discussed.

- Shewhart control charts, including individuals (I) charts and means (Xbar) charts for the average level of the bioassay, and moving range (MR), range (R) and standard deviation (S) charts for the variability of the bioassay, are introduced.

- CUSUM and exponentially weighted moving average (EWMA) charts are also briefly described.

- Sets of rules for monitoring the process via the charts are presented and the choice of an appropriate subset of these is discussed.

- Actions that may be required following a breach of rules, for example the detection of a trend in a parameter, are considered.

10

Bioassay updates and modifications

Over the lifecycle of a bioassay, changes become necessary. The reference standard lot may expire or run out; the same goes for other reagent lots and test kit lots. Cell banks may need to be replaced. In any of these situations, the continued validity of the bioassay needs to be checked.

A new reference standard lot will not have exactly the same potency as its predecessor, though it may be 'close enough'. Any change, though, will shift the distribution of bioassay-reported potencies for a manufacturing process and hence will change the analytical profile and the probability of an out-of-specification result unless corrective action is taken. It is therefore vital to quantify the potency of the new standard so that allowance can be made for the change.

Another change that needs to be addressed carefully is the transfer of a bioassay from one laboratory to another. The accuracy and precision of the bioassay when conducted at the new laboratory will not be identical to those which were evaluated at the original laboratory; assurance is needed that the bioassay is still performing adequately.

This chapter covers the statistical aspects associated with ensuring that changes in a bioassay do not result in unacceptable performance.

10.1   Characterising the reference standard

In bioassays, the reference standard is usually a representative batch from the manufacturing process that has acceptable clinical properties. In some cases, the reference standard may have an assigned absolute potency, in terms of concentration or units of activity. The absolute potency of a test sample can be calculated by multiplying its relative potency by the assigned potency of the reference standard. The specification limits (for batch release or for stability) may be in terms of concentration or relative potency.

As in previous chapters, we will use the subscripts S and T to represent the reference standard and test sample, respectively. We will use the superscripts $(\cdot)^{(O)}$ to indicate if a quantity has been measured on or against the original standard, and $(\cdot)^{(N)}$ to indicate if a quantity is measured on or against a new (replacement) standard. For example, the relative potency for a test batch, relative to the original reference batch, will be denoted as $RP^{(O)}$.

When making a decision for batch release, it is the quantity measured against the original standard that is most applicable.

If the original reference batch has an assigned, concentration, $C_S^{(O)}$, then the concentration for the test sample is given as:

$$C_T = RP^{(O)} \times C_S^{(O)}.$$

If the reference standard does not have an assigned concentration, then the reported result for a test sample will be the relative potency, $RP^{(O)}$; in this case the specification limits will apply to the relative potency and will only be relevant to the potency relative to the original reference standard.

The introduction of a new reference standard impacts both the calculation of C_T and the calculation of $RP^{(O)}$. Let the relative potency for a test batch relative to the new reference standard be denoted as $RP^{(N)}$. Then, the concentration for the test batch will be given by:

$$C_T = RP^{(N)} \times C_S^{(N)},$$

where $C_S^{(N)}$ is the concentration of the new reference standard. The concentration of the new reference standard must be precisely evaluated by testing the new standard as a test sample against the original standard. We refer to this relative potency as the 'offset' and denote it as $Off^{(N:O)}$. Then, the concentration of the new reference standard can be found as:

$$C_S^{(N)} = Off^{(N:O)} \times C_S^{(O)}.$$

Therefore, the concentration for a new test batch is:

$$C_T = RP^{(N)} \times Off^{(N:O)} \times C_S^{(O)}.$$

If the specification is set in terms of relative potency, then the test sample's potency, relative to the original reference standard, $RP^{(O)}$, is required. This must be estimated via its potency $RP^{(N)}$ and the offset:

$$RP^{(O)} = RP^{(N)} \times Off^{(N:O)}.$$

Figure 10.1 shows an example where a test sample has been tested against two reference standard batches: the original standard is shown in dark grey and the new standard in black. The new standard curve is to the right of the original standard curve, indicating that a higher dose is required to achieve the same response. Therefore, the estimate of the offset in this case is less than 1. The potency of the test sample relative to the new reference standard will therefore be higher than it was relative to the original reference standard.

In the example, $RP^{(N)} = 0.85$. The estimate of the offset is $Off^{(N:O)} = 0.93$. Therefore, the relative potency of the test sample, relative to the original standard, can be found as:

$$
\begin{aligned}
RP^{(O)} &= RP^{(N)} \times Off^{(N:O)} \\
&= 0.85 \times 0.93 \\
&= 0.79.
\end{aligned}
$$

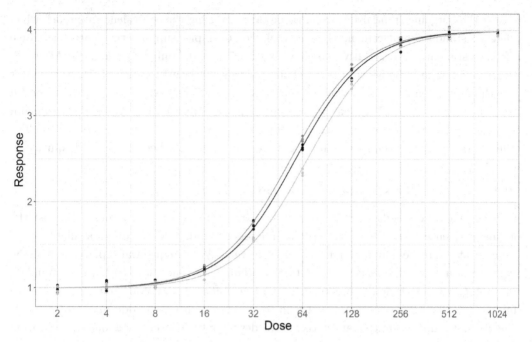

Sample — New Standard — Original Standard — Test Sample

FIGURE 10.1

Dose-response relationships for a test sample and the original and new reference standards. Since the curve for the new reference standard sits to the right of the original standard, the relative potency of the test sample will be higher when tested against the new standard.

Since the new standard is less potent than the original, the potency of a test sample relative to the new standard will be inflated. The offset is applied to adjust the potency to counteract this effect.

The offset, or the potency of the new reference standard relative to the original reference standard, is fundamental to the calculation of potency for a test sample following the replacement of the reference standard. However, there will be uncertainty when estimating the value of this offset. The uncertainty in the estimate of the offset will further add to the uncertainty in the reported results for the bioassay. Therefore, it will adversely impact on the bioassay performance in terms of precision.

The number of repeat evaluations for determining the offset with enough precision will generally be higher than the number required for the release of a batch of material. If the offset value is found to be close to 1, then it may be suitable to report the results without correcting for the difference between the original and new reference standards. This could be done by conducting an equivalence test, where equivalence is demonstrated if the confidence interval for the offset (on the log scale) falls within equivalence bounds $(-\delta, \delta)$.

Ideally, aliquots of the original reference standard lot would be retained for assessment of comparability with all later reference standards. However, if this is not possible then the reference standards will need to be replaced sequentially by testing each new standard against the outgoing standard. Care must be taken in using the equivalence approach when reference standard replacement is sequential. For example, suppose that each successive reference standard has a relative potency that is 97% of the outgoing standard. After one replacement, it may be acceptable to claim equivalence with the original. However, after only three replacements the potency of the latest standard relative to the original standard would be $97\%^3 = 91\%$. This may be too far from 100% of the original standard to claim equivalence. For each reference standard replacement the offset should be recorded to allow a link to the original, even if some intermediate standards have been considered equivalent.

10.1.1 Reference standard qualification study

The new reference standard must be tested several times in the assay, ideally against the original reference standard. If the original reference standard is no longer available, then the current reference standard must be used instead. With sequential replacement of the reference standard, the offset relative to the original reference standard will accumulate uncertainty with each new estimate. That is, the confidence interval around the offset to link back to the original standard will tend to get larger with each additional replacement. Therefore, it is important to plan for this in the design of the reference standard qualification studies to ensure that each offset is estimated with sufficient precision.

10.1.1.1 Number of runs required

We want to estimate the offset, i.e., the RP of the new reference lot, precisely. In other words, the 95% confidence interval for the estimate of the offset must be narrow. The width of the confidence interval will depend on the number of runs and the standard deviation of repeated measures of the same relative potency, or, equivalently, the intermediate precision of the assay.

Suppose the reference standard qualification study includes n sessions, each of k repeats. The new reference standard is simply a test sample at this point. The variance of the estimate of the offset on the log scale, $\log \mathrm{Off}^{(\mathrm{N:O})}$ will be:

$$\sigma^2_{\log \mathrm{Off}^{(\mathrm{N:O})}} = \frac{\sigma^2_{\text{Between session}}}{n} + \frac{\sigma^2_{\text{Within session}}}{kn}.$$

The 95% confidence interval for the offset, on the log scale, can be written as:

$$\widehat{\log \mathrm{Off}}^{(\mathrm{N:O})} \pm t_{0.975, df} \times \sqrt{\frac{\widehat{\sigma}^2_{\text{Between session}}}{n} + \frac{\widehat{\sigma}^2_{\text{Within session}}}{kn}}, \qquad (10.1)$$

where $\widehat{\log \mathrm{Off}}^{\mathrm{N:O}}$, $\widehat{\sigma}^2_{\text{Between session}}$ and $\widehat{\sigma}^2_{\text{Within session}}$ are estimated from the qualification study data. The degrees of freedom for the t-statistic will be given by $df = (n-1)$ if only one repeat is tested per session and $k = 1$, or can be approximated using the Satterthwaite approach if $k > 1$ (see Section 3.4.2). On the original scale, the lower and upper 95% confidence limits for the offset are:

$$\frac{\widehat{\mathrm{Off}}^{(\mathrm{N:O})}}{\mathrm{antilog}\left(t_{0.975, df} \times \sqrt{\frac{\widehat{\sigma}^2_{\text{Between session}}}{n} + \frac{\widehat{\sigma}^2_{\text{Within session}}}{kn}}\right)}, \qquad (10.2)$$

TABLE 10.1

Multiplier, M, for the 95% confidence limits for the offset $\mathrm{Off}^{(\mathrm{N:O})}$ assuming $k = 1$.

Number of	Intermediate precision		
sessions	5%	10%	20%
3	1.129	1.267	1.573
4	1.081	1.164	1.337
5	1.062	1.126	1.254
6	1.053	1.105	1.211
7	1.046	1.092	1.184
8	1.042	1.083	1.165
9	1.038	1.076	1.150
10	1.036	1.071	1.139
15	1.027	1.054	1.106
20	1.023	1.046	1.089
25	1.020	1.040	1.078
30	1.018	1.036	1.070

and

$$
\widehat{\mathrm{Off}}^{(\mathrm{N:O})} \times \mathrm{antilog}\left(t_{0.975,df} \times \sqrt{\frac{\widehat{\sigma}^2_{\text{Between session}}}{n} + \frac{\widehat{\sigma}^2_{\text{Within session}}}{kn}} \right). \tag{10.3}
$$

Alternatively, we can express these as:

$$
\left(\frac{\widehat{\mathrm{Off}}^{(\mathrm{N:O})}}{M}, \ \widehat{\mathrm{Off}}^{(\mathrm{N:O})} \times M \right),
$$

where M is a multiplier (original scale) that is related to the width of the confidence interval.

One way to ensure that there is sufficient precision is to place limits on the width of the confidence interval for the estimate of the offset; or equivalently ensure that the multiplier M is not too large. The sample size for the reference qualification study can then be calculated for a given confidence interval width. One option is to assume that the bioassay precision has not changed and to use the precision estimates from the validation study. If, following the laboratory work, the confidence interval is found to be wider than intended, additional data can be collected (and pooled with the original data). This is because a confidence interval based on a larger sample size will tend to be narrower. This flexibility should be built into the reference standard qualification protocol. Another method for planning a reference qualification study is to use a metric known as the critical fold difference; we will discuss this further in the next section.

Table 10.1 shows the multiplier for the confidence limits for various combinations of the number of sessions and intermediate precision, assuming $k = 1$ repeat per session. The multiplier initially decreases rapidly as the number of runs increases, with the return diminishing as the sample size gets larger.

The critical fold difference

The critical fold difference (CFD) between the potencies of the original and new reference standards can be defined as:

$$\text{CFD} = \text{antilog}\left(2 \times \sqrt{\frac{\sigma^2_{\text{Between session}}}{n} + \frac{\sigma^2_{\text{Within session}}}{kn}}\right)$$
$$= \text{antilog}\left(2 \times \sqrt{\sigma^2_{\log \text{O}^{(\text{N:O})}}}\right).$$

The CFD is an approximation to the multiplier in Equations (10.2) and (10.3) and can be used to characterise the width of the confidence interval. The CFD is an approximation as it replaces the value of the t-statistic with a value of 2; as such, it does not reflect the additional uncertainty accounted for by the t-distribution when the amount of information to estimate the variance is small (i.e., when the degrees of freedom are small). When the degrees of freedom associated with the t-statistic are small, the discrepancy between the t-statistic and the approximated value of 2 will be large. For example, at $df = 5$ the value of $t_{0.975,5} = 2.57$. However, as the degrees of freedom increase the approximation will become closer and at $df = 60$, $t_{0.975,60} = 2.00$; a sample size this large may not be applicable in the context of bioassay. For $df > 60.44$, the value of the t-statistic will be less than 2; its minimum value is $t_{0.975,\infty} = 1.96$.

The simplicity of the CFD formula can make it useful when considering the design of reference qualification studies with multiple replicates per session without the need to estimate the degrees of freedom using complex formulas. To design a study with a target CFD in mind we need initial estimates of $\sigma^2_{\text{Between session}}$ and $\sigma^2_{\text{Within session}}$ to get started. These often come from the intermediate precision assessment carried out during the validation study.

Table 10.2 and Figure 10.2 show the critical fold difference for values of $k = 1$ and 3 combined with values of n from 1 to 20 for $\sigma^2_{\text{Between run}} = 0.0036$ and $\sigma^2_{\text{Within run}} = 0.0018$; this is equivalent to a bioassay with an intermediate precision of 7.6%. As for the multiplier for the upper confidence limit, CFD initially decreases rapidly as the number of runs increases, with diminishing returns once we reach about 10 runs. The number of repeats within run has less of an impact in this case because the within-run variance is much smaller than the between-run variance. For example, with 1 result per run and 10 runs, the upper 95% confidence limit for the offset is 4.8% above the estimate. If the target is to estimate the offset to within 5% then this would be a good solution. An alternative would be to include 7 runs, each including 3 repeats.

Example: reference qualification study

Suppose that a reference qualification study is being planned for an assay with an intermediate precision of 7.6%, where $k = 1$ repeat per session will be measured. If we desire an upper 95% confidence limit no greater than 105% of the estimate and a lower confidence limit no lower than $\frac{1}{1.05} = 95.2\%$ then we require a sample size of 12 sessions. Note that this is larger than the 9 sessions given in Table 10.2 as this calculation was done using $t_{0.95,11} = 2.20$ instead of a value of 2 used for the CFD.

TABLE 10.2

Critical fold difference for 1–20 sessions, for $k = 1$ and $k = 3$ repeats per session and $\sigma^2_{\text{Between run}} = 0.0036$ and $\sigma^2_{\text{Within run}} = 0.0018$; this is equivalent to a bioassay with an intermediate precision of 7.6%.

Sessions $= n$	$k = 1$ repeat per session	$k = 3$ repeats per session
1	1.158	1.138
2	1.110	1.096
3	1.089	1.078
4	1.076	1.067
5	1.068	1.060
6	1.062	1.054
7	1.057	1.050
8	1.053	1.047
9	1.050	1.044
10	1.048	1.042
11	1.045	1.040
12	1.043	1.038
13	1.042	1.037
14	1.040	1.035
15	1.039	1.034
16	1.037	1.033
17	1.036	1.032
18	1.035	1.031
19	1.034	1.030
20	1.033	1.029

TABLE 10.3

Relative potency results for reference qualification study.

Session	RP	log RP
1	0.93	−0.073
2	0.86	−0.151
3	0.86	−0.151
4	1.20	0.182
5	0.87	−0.139
6	0.97	−0.030
7	0.98	−0.020
8	1.08	0.077
9	1.00	0.000
10	0.89	−0.117
11	0.95	−0.051
12	1.09	0.086

Table 10.3 presents the relative potency estimates for 12 sessions, while Table 10.4 presents the summary statistics for these data. It can be seen that the geometric mean relative potency among the 12 results is 0.968 with a 95% confidence interval of (0.906, 1.035). This confidence interval has lower and upper 0.935 and 1.069 times the estimate.

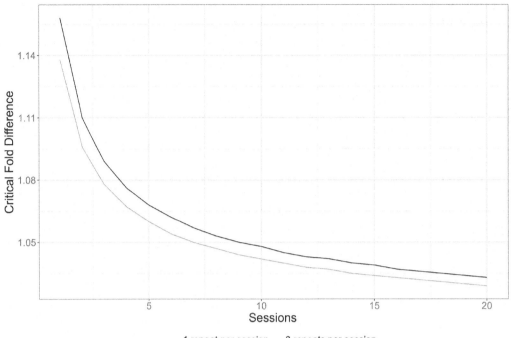

FIGURE 10.2

Critical fold difference for 1–20 sessions, $k = 1$ and $k = 3$ repeats per session, $\sigma^2_{\text{Between run}} = 0.0036$ and $\sigma^2_{\text{Within run}} = 0.0018$. Initially the CFD drops quickly as the number of sessions is increased, but there are diminishing returns for large values of n.

TABLE 10.4

Summary results for reference qualification study.

	Log scale	Original scale
n	12	
Mean	−0.032	0.968
Standard deviation	0.105	11.1%
$t_{0.975,(11)}$	2.201	
Estimate of offset and 95% CI	−0.032 (−0.099, 0.035)	96.8% (90.6%, 103.5%)

The intermediate precision for this study was 11.1% and exceeded the 7.6% for which this study was planned. As such, the half width for the confidence interval and confidence limits were wider than desired. However, additional data can be collected so that the confidence interval can be estimated with greater precision. For an intermediate precision of 11.1% an additional 9 sessions, for a total of 21 results, should be sufficient based on Formula 10.3.

10.1.1.2 Precision required for the estimate of the offset

In the example, the aim was to estimate the offset to within 5%. We now examine how to set this target for the precision of the offset.

The estimate of log RP against the original reference standard, for a test sample which has been tested against the new reference standard is:

$$\widehat{\log \text{RP}}^{(\text{O})} = \widehat{\log \text{RP}}^{(\text{N})} + \widehat{\log \text{Off}}^{(\text{N}:\text{O})}.$$

If we anticipate that repeated replacements of the reference standard will be made by testing the new reference standard against the previous one, rather than reverting to the original, then a new offset will need to be calculated on each occasion. Suppose that there will be R replacements of the reference standard, indexed as $\text{N}_1 \ldots \text{N}_R$. Then, the RP for a test sample measured against the latest reference standard, N_R, will be adjusted to provide its RP against the original reference standard as follows:

$$\widehat{\log \text{RP}}^{(\text{O})} = \widehat{\log \text{RP}}^{\text{N}_R} + \widehat{\log \text{Off}}^{(\text{N}_R:\text{N}_{R-1})} + \cdots + \widehat{\log \text{Off}}^{(\text{N}_2:\text{N}_1)} + \widehat{\log \text{Off}}^{(\text{N}_1:\text{O})}. \quad (10.4)$$

The variance of the estimate is the sum of the variances of the terms on the right of Equation (10.4). That is, the variance of the RP estimate (on the log scale) will be:

$$\sigma^2_{\log \text{RP}(\text{O})} = \sigma^2_{\log \text{RP}(\text{N}_R)} + \sum_{i=2}^{R} \sigma^2_{\log \text{Off}(\text{N}_i:\text{N}_{i-1})} + \sigma^2_{\log \text{Off}(\text{N}_1:\text{O})}.$$

If we assume that the bioassay precision itself does not change (that is, the new reference standards have not affected the precision of the bioassay itself), the variance of $\widehat{\log \text{RP}}^{(\text{N}_R)}$ will be as established in the validation study:

$$\sigma^2_{\log \text{RP}(\text{N}_R)} = \sigma^2_{\text{Between session}} + \sigma^2_{\text{Within session}}.$$

The variance of $\widehat{\log \text{RP}}^{(\text{O})}$ is therefore:

$$\sigma^2_{\log \text{RP}(\text{O})} = \sigma^2_{\text{Between session}} + \sigma^2_{\text{Within session}} + R \times \sigma^2_{\log \text{Off}(\text{N}_1:\text{O})},$$

where $\sigma^2_{\log \text{Off}(\text{N}_1:\text{O})}$ is the variance of $\widehat{\log \text{Off}}^{(\text{N}_1:\text{O})}$ from the original reference standard qualification study, assumed to be constant across all replacement studies, and R replacements have been conducted.

For example, suppose the ATP specifies that the IP must be no more than 10%. Also suppose that the bioassay validation study has demonstrated that the IP is 8%. In terms of the variance of $\widehat{\log \text{RP}}^{(\text{O})}$, the ATP implies that:

$$\sigma^2_{\log \text{RP}(\text{O})} \leq \left[\log \left(\frac{10}{100} + 1 \right) \right]^2,$$

and the validation study result implies that:

$$\sigma^2_{\text{Between session}} + \sigma^2_{\text{Within session}} \leq \left[\log \left(\frac{8}{100} + 1 \right) \right]^2.$$

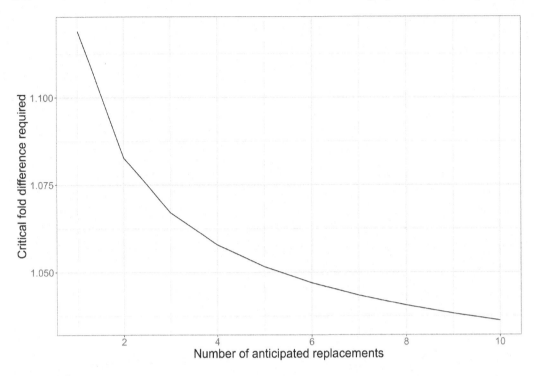

FIGURE 10.3
Critical fold difference required for the number of planned replacements, R, ranging from 1 to 10, when the ATP specifies that the intermediate precision GCV must be no more than 10%.

Therefore:

$$R \times \sigma^2_{\log \text{Off}^{(\text{N}_1:\text{O})}} \leq \left[\log\left(\frac{10}{100} + 1\right)\right]^2 - \left[\log\left(\frac{8}{100} + 1\right)\right]^2$$
$$= 0.0032$$

ensures that the ATP will still be met even when the estimated offset for the reference standard change is applied. This can be expressed as:

$$\hat{\sigma}^2_{\log \text{Off}^{(\text{N}_1:\text{O})}} \leq \frac{0.0032}{R},$$

or:

$$CFD \leq \text{antilog}(2 \times \sqrt{0.0032/R}).$$

Figure 10.3 and Table 10.5 show the critical fold difference required for R ranging from 1 to 10, when the ATP specifies that the intermediate precision must be no more than 10% and the bioassay validation study has demonstrated that the intermediate precision is 8%. The required CFD initially decreases rapidly as the number of anticipated replacements of the reference standard increases. This is because extra variability is introduced into the potency estimate by each estimated offset. The lower the required CFD, the more repeats will be needed for the evaluation of each offset.

TABLE 10.5

Critical fold difference required for the number of planned replacements, R, ranging from 1 to 10, when the ATP specifies that the IP must be no more than 10%.

Number of anticipated reference standard replacements = R	CFD required for each offset
1	1.119
2	1.083
3	1.067
4	1.058
5	1.052
6	1.047
7	1.043
8	1.041
9	1.038
10	1.036

If only one replacement is anticipated, then in this situation (ATP precision requirement 10%, actual precision 8%), the CFD needs to be at most 1.119 to ensure that the ATP is still met. If at each replacement the new reference standard is to be compared directly to the original batch, then this requirement still holds for each new reference standard batch regardless of how many are anticipated. However, if the new reference standard batches are compared only with their predecessor, the CFD requirement is much more stringent. With 4 replacements the CFD requirement is 1.058.

10.1.2   Other considerations

In addition to estimating the offset, when changing reference standards there are other considerations that to be made. If the potency of the new reference standard is different from the original, then, in the case of a 4PL model, its EC_{50} will also be different. In the case of a straight-line model, the intercept will be different. Therefore, if there are system suitability criteria involving these parameters, the limits will need to be reviewed. If there is a criterion for the relative potency of the assay control, then this may also need to be updated. If the assay control was and continues to be an independent replicate of the reference standard and the criterion is centred around a value of 1, then an update may not be necessary.

10.2   Assay transfer

Transferring an assay to a different laboratory involves many practical issues, and, once set up at the new location, steps must be taken to ensure that the assay is still meeting its ATP. Careful examination of the available data prior to transfer, and incorporating the available information into the transfer plan from a statistical point of view, can help to minimise the experimental work required.

Transfer of analytical procedures, also known as TAP, or method transfer, is the documented process that qualifies a laboratory to use an analytical test procedure that originated in another laboratory. The idea is to ensure that that the new laboratory (or the 'receiving unit', RU) has both the procedural knowledge and ability to perform the transferred analytical procedure as intended by the originating laboratory (the transferring unit or sending unit, SU).

Several approaches are possible and include comparative testing, co-validation and complete (or partial) validation at the receiving unit [61]. We further define each of these approaches below.

Comparative testing: Homogeneous lots of the target material from standard production batches or samples intentionally prepared for the test (e.g., by spiking relevant accurate amounts of known impurities into samples) are each tested at both the SU and the RU and the results are compared.

Co-validation: The laboratory that performs the validation of an analytical procedure is then qualified to run the procedure. Therefore, if both the sending unit and the receiving unit are involved in the validation, as part of a documented inter-laboratory collaboration, the receiving unit will automatically be qualified to run the procedure [61].

Complete / partial validation at the receiving unit: Partial validation has been defined [5] as the demonstration of assay reliability following a modification of an existing bioanalytical method that has previously been fully validated. The nature of the modification will determine the extent of validation required.

We describe the statistical aspects of comparative testing in this chapter.

10.2.1 Comparative testing

It is acceptable to use a single lot for the transfer [61], because the aim is the evaluation of the analytical procedure's performance at the receiving site, rather than the manufacturing process as a whole. However, several lots with varying potencies are often included so that the procedure's performance at the receiving site across the range can be evaluated.

There are two aims of a transfer study. The first aim is to verify that the results for a lot are comparable / equivalent at the two sites, while the second is to ensure that the precision at the RU is as good as it is at the SU (or is at least acceptable).

10.2.1.1 Comparability/equivalence of the results

We will begin with the assumption that a single lot is to be tested multiple times at each of the sending and receiving units. We will use the subscripts 1 and 2 to represent the sending and receiving unit, respectively. Then let $i = 1, \ldots, n_1$ index the results at the sending unit and $j = 1, \ldots, n_2$ index the results at the receiving unit; note that it is not necessary that the number of replicates be the same. Recall that the results are normally distributed on the log scale and so all calculations will be conducted first on the log scale and then back transformed. We define the ith observation at the sending unit as $\log(\text{RP}_{1i})$ and the

j^{th} observation at the receiving unit be denoted as $\log(RP_{2j})$. For the purpose of planning the transfer study, in the absence of information about the variance of the results at the RU we assume that this is the same as seen at the SU. Therefore, the results will be have distributions as follows:

$$\log(RP_{1i}) \sim N\left(\mu_1, \sigma^2\right),$$

$$\log(RP_{2j}) \sim N\left(\mu_2, \sigma^2\right).$$

The aim of the comparative testing is to show that the mean values are equivalent. Therefore, the null hypothesis is:

$$H_0:\ |\mu_1 - \mu_2| \geq \delta,$$

and the alterative hypothesis is:

$$H_A:\ |\mu_1 - \mu_2| < \delta.$$

If the 90% confidence interval for $\mu_1 - \mu_2$ lies within $(-\delta, \delta)$ then equivalence, or comparability, will have been demonstrated.

Let $\overline{\log(RP_1)}$ and s_1 be the mean and standard deviation of the log-transformed relative potencies at the sending unit. The analogous quantities for the receiving unit can be defined by replacing the subscript 1 with 2. Then in the case of a single lot, the 90% confidence interval for the difference is given by:

$$\left[\overline{\log(RP_1)} - \overline{\log(RP_2)}\right] \pm t_{0.95,(n_1+n_2-2)} \times s_p \times \sqrt{\frac{1}{n_1} + \frac{1}{n_2}},$$

where s_p is the pooled standard deviation, given by:

$$s_p = \sqrt{\frac{(n_1 - 1)s_1^2 + (n_2 - 1)s_2^2}{n_1 + n_2 - 2}}.$$

The confidence interval for the geometric mean ratio between the sending and receiving units can be obtained by taking the antilog of the confidence limits.

10.2.1.2 Multiple lots

When multiple lots are tested, it is common to assume that the difference between the sending and receiving units is constant across the lots, but that the lots themselves might have different relative potencies. To conduct this type of analysis, we can use a multiple regression model that includes an effect for each lot and a site effect (sending vs receiving unit). Equivalently, a two-way analysis of variance (ANOVA) with factors for site and lot can be performed. The two-way ANOVA will fit a separate mean to each lot, and estimate a common difference between the sending and receiving units for each lot. While it is possible to allow for the effect of the site to depend on the lot, this makes it difficult to generalise the results to future (untested) lots.

Let $i = 1, \ldots, n$ index over the combined set results for all lots from both units, and let $l = 1, \ldots, L$ index the lots. We will assume that the first lot, $l = 1$, and site 1, the sending unit, are the reference levels of the regression model. Then, the multiple linear regression model is given by:

$$\log \mathrm{RP}_i = \beta_0 + \sum_{l=2}^{L} \beta_{\mathrm{lot},l} \times \mathbb{I}(\mathrm{lot}_i = l) + \beta_{\mathrm{Site}} \times \mathbb{I}(\mathrm{site}_i = 2) + \epsilon_i, \qquad (10.5)$$

where:

- β_0 is the intercept;

- $\beta_{\mathrm{lot},l}, l = 2, \ldots, L$ is the mean difference between first lot and lot l (on the log scale);

- $\mathbb{I}(\mathrm{lot}_i = l)$ is an indicator that takes on a value of 1 if the ith belongs to lot l, and 0 otherwise;

- β_{Site} is the mean difference between the measured potency at the receiving unit, compared to the sending unit (on the log scale);

- $\mathbb{I}(\mathrm{site}_i = 2)$ is an indicator that takes on value of 1 if the ith observation belongs to site 2 (the receiving unit);

- ϵ_i is a random error term with mean 0 and variance σ^2.

In this case, demonstrating equivalence involves showing that the site effect measured by β_{Site} is negligible. Therefore, the relevant null hypothesis is:

$$H_0\colon |\beta_{\mathrm{Site}}| \geq \delta,$$

and the alternative hypothesis is:

$$H_{\mathrm{A}}\colon |\beta_{\mathrm{Site}}| < \delta.$$

If a 90% confidence interval for β_{Site} lies within $(-\delta, \delta)$ then equivalence, or comparability, will have been demonstrated.

10.2.1.3   Choice of equivalence bounds

When designing a transfer study, the value for the equivalence bounds for the mean difference between the sending and receiving units (on the log scale), δ, is often set arbitrarily: e.g. $\delta = \ln(1.25) = 0.2231$. However, a limit that is driven by the ATP is more meaningful. For example, suppose the ATP specifies that the relative bias must be no more than 12%: that is, RB must lie within $(-11\%, 12\%)$. If we assume that the true bias at the SU is zero then meaningful limits for the comparability coincide with those for the RB: $\delta = \ln(1.12) = 0.1133$. If the difference between the means sits at the limit, δ, then the RB at the RU is at the limit of acceptability.

If the bioassay validation study has demonstrated that the relative bias at the SU is 5%, then for the relative bias to lie within $(-11\%, 12\%)$ at the RU, $(\mu_1 - \mu_2)$ must lie between $[\log(0.89) - \log(1.05)] = -0.1653$ and $[\log(1.12) - \log(1.05)] = 0.0645$. This is equivalent to the ratio of the two geometric mean RP values lying between approximately 0.85 and 1.07. The interval is asymmetric because of the existing positive bias at the sending unit: there is more 'headroom' at the lower end of the interval.

TABLE 10.6

Sample size required for 90% power to demonstrate equivalence for different limits and intermediate precision. Sample size calculations assume that only a single lot is being tested multiple times.

		Assumed intermediate precision					
Limits		5%		10%		20%	
log scale = δ	Back-transformed	Per site	Total	Per site	Total	Per site	Total
±0.2231	(0.800, 1.250)	3	6	5	10	16	32
±0.0953	(0.909, 1.100)	7	14	23	46	80	160
±0.0488	(0.952, 1.050)	23	46	84	168	303	606

10.2.1.4 Number of runs required for comparability

The design of a transfer protocol involves calculating the required number of runs. Even if the observed difference between the unit means is zero, the confidence interval needs to be adequately narrow for comparability to have been demonstrated. For a given value of δ, and assumed value of s, the minimum values of n_1 and n_2 for the L lots can be determined such that, if the difference between that two units is truly zero, there is a high chance that study will demonstrate comparability. This chance is called the power of the study.

Standard statistical software is available for this calculation [49]. Table 10.6 gives the sample size for the number of independent results at each site (and in total) for different values for the equivalence bounds and intermediate precision. If multiple lots are tested, an extra parameter will be estimated per additional lot. Conservatively, an additional sample should be added to the sample size for each additional lot.

10.2.1.5 Acceptable variability at the RU

The precision achieved at the RU can be calculated as in a validation study, i.e. as the IP, and compared with the validated value from the SU. The IP (or its upper 95% one-sided confidence limit) must fall below the pre-set limit. The IP (and its upper 95% one-sided confidence limit) is calculated as described in Chapter 8. Also see Chapter 8 for calculating the sample size based on the IP.

10.2.1.6 Examples: assay transfer study

One lot

Suppose an assay transfer study is to be conducted, and we assume that the limits for comparability are (91%, 110%) and the IP is 5%. Then a minimum sample size of $n = 7$ replicate measurements are required for 90% power (see Table 10.6). If the limit for the IP is 12%, then a minimum sample size of 8 is required for 90% power. Suppose, a decision was made to run 9 assays.

Table 10.7 displays the relative potency and corresponding log-transformed values for nine measurements of a single lot at each of the sending and receiving units. Table 10.8 shows the summary statistics for the transfer study. The estimated mean difference between

TABLE 10.7

Relative potency and corresponding log-transformed values for nine measurements of a single lot at each of the sending and receiving units.

Sending unit		Receiving unit	
RP	log(RP)	RP	log(RP)
0.98	−0.02	0.96	−0.04
0.91	−0.09	1.03	0.03
0.94	−0.06	1.15	0.14
0.96	−0.04	0.99	−0.01
1.02	0.02	0.96	−0.04
1.02	0.02	0.96	−0.04
0.94	−0.06	1.11	0.10
0.98	−0.02	1.04	0.04
0.92	−0.08	0.99	−0.01

TABLE 10.8

Summary statistics for an assay transfer based on multiple repeat tests of a single lot.

	Sending unit	Receiving unit
n	9	9
Mean	−0.0381	0.0189
Standard deviation	0.0414	0.0660
$t_{0.95,(16)}$		1.734
Difference (90% CI), log scale	−0.0570 (−0.1024, −0.0117)	
Original scale		94% (90%, 99%)
IP (upper 95% confidence limit)		7% (12%)

the sending and receiving units (log scale) was −0.0570 with a 90% confidence interval of (−0.1024, −0.0117). On the original scale, this indicates that the results at the receiving unit, on average are 94% of those at the sending unit, with a 95% confidence interval (for the multiplicative difference) of (90%, 99%). In this case, the lower limit is smaller than limit for comparability and comparability is not demonstrated. For the IP, the estimate at the RU was 7% with a 95% upper confidence limit of 12%. For this study, the criterion for the variability was met.

Three lots

In this second example, we demonstrate the analysis required when more than 1 lot will be tested. We only focus on comparability here as the analysis for IP is demonstrated in Chapter 8. Suppose an assay transfer study is to be conducted, and we assume that the limits for comparability are (80%, 125%) and the IP is 20%. Then a minimum sample size of $n = 16$ replicate measurements are required for 90% power (see Table 10.6) if only one lot were to be tested. For a study with three lots, a sample size of 18 per site is required. As the amount of material was limited, a risk was taken and a sample size of 15 was selected.

TABLE 10.9

Relative potency and corresponding log-transformed values for three lots with five measurements each at each of the sending and receiving units.

Lot A				Lot B				Lot C			
Sending		Receiving		Sending		Receiving		Sending		Receiving	
RP	log RP	RP	log RP	RP	log RP	RP	log RP	RP	log RP	RP	log RP
0.92	−0.083	0.95	−0.051	1.03	0.030	1.04	0.039	1.21	0.191	1.19	0.174
0.89	−0.117	0.88	−0.128	1.04	0.039	0.99	−0.010	1.21	0.191	1.23	0.207
0.95	−0.051	0.93	−0.073	1.02	0.020	1.15	0.140	1.16	0.148	1.28	0.247
0.84	−0.174	0.90	−0.105	1.09	0.086	1.01	0.010	1.18	0.166	1.22	0.199
0.94	−0.062	0.91	−0.094	0.95	−0.051	0.99	−0.010	1.11	0.104	1.20	0.182

TABLE 10.10

Summary statistics for an assay transfer based on testing three lots.

	Estimate	Std. error
Intercept	−0.1036	0.0159
Lot B	0.1231	0.0194
Lot C	0.2747	0.0194
Site = Receiving unit	0.0194	0.0159
90% CI for β_{Site}, log scale	(−0.00767, 0.04646)	
90% CI for β_{Site}, original scale	(99%, 105%)	

Table 10.10 displays the relative potency and corresponding log-transformed values for three lots with five measurements each at each of the sending and receiving units. These data are then plotted in Figure 10.4. From the plot there is a clear difference in the relative potency for each lot but no obvious difference in the measurements between the two sites.

Table 10.10 displays the regression parameter estimates for the model in Equation (10.5). The estimate for $\hat{\beta}_{\text{site}} = 0.0194$, with a 90% CI on the log scale of $(−0.00767, 0.04646)$. When back transformed by taking the antilog, the estimate for the geometric mean ratio between sites is 102% with a confidence interval of $(99\%, 105\%)$. This indicates that the results at the receiving unit are, on average, slightly higher than those obtained by the sending unit. In this case, the confidence interval for the difference between the sites falls well inside the comparability limits and the study passes for comparability.

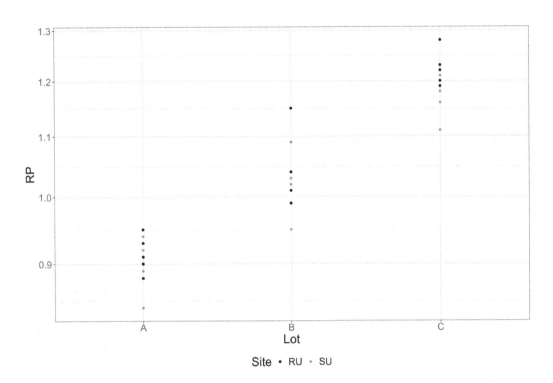

FIGURE 10.4

Relative potency (plotted on log scale) for three lots, for each of the sending and receiving units. There is a clear difference in the relative potency for each lot but no obvious difference in the measurements between the two sites.

10.3 Chapter summary

- Changes in the bioassay that are likely to be encountered over the course of its lifetime are considered, including the transfer of a bioassay method from one laboratory to another and the introduction of a new reference standard.

- Acceptance criteria for assurance that the two laboratories are performing comparably must be developed, including the average difference between, and the variability of, the laboratories' results.

- Methods for the design of an assay transfer study, including the numbers of runs needed at each laboratory, and the analysis of the results, are presented.

- For the introduction of a new reference standard, the potency of the candidate lot needs to be compared with that of the current reference standard, or, ideally, directly with that of the original reference standard.

- A study must be designed to allow the estimation of the relative potency of the candidate lot to be made with adequate precision. A useful measure, the critical fold difference, is explained.

- Variance components analysis is key to these processes.

Part IV

Appendices

A

Statistical/mathematical background

In this Appendix we provide a background on basic introductory statistics. Most of the content presented here would be covered in a first course in statistics at a university level. There are many introductory statistics textbooks that cover this material in far more detail than we can cover here; see [13, 43, 65]. This Appendix will draw on examples from outside the bioassay context. We then apply these concepts within the bioassay context throughout the main chapters of this textbook.

A.1 Introduction

Statistical concepts and methods are critical to allowing informed decisions to be made in the presence of variability. For example, characterising the relative potency of a new test sample using an analytical method that is variable. It is often difficult to understand or detect patterns in large data sets and summarisation of the data is required to provide interpretable information.

A common statistical problem is to estimate descriptive characteristics of the population known as parameters. Common population parameters include measures of central tendency such as the mean or proportion, measures of variability such as the variance and standard deviation, as well as measures of association such as the correlation and regression parameters.

An example of a population parameter would be the true mean income for all individuals living in a particular region, say Scotland. Typically sampling the entire population is not feasible, and so estimates are instead based on a sample (or subset/portion) of the population. A statistic is a descriptive measure for a sample and is used to estimate the corresponding population parameter. For example, the sample mean income for a sample of 100 individuals living in Scotland could serve as an estimate of the population mean.

DOI: 10.1201/9781003449195-A

A.2 Variables and variable types

A variable is any characteristic that varies from one individual or item in a study to the next. Variables can be split into two broad categories: quantitative and qualitative. It is important to consider how the data are to be collected before beginning a study as it will determine the statistical procedures that can be applied and the resulting conclusions that can be drawn.

Quantitative variables have a numeric value, for example, the height or weight of an individual, the time it takes to travel through airport security, or the relative potency of a new product lot. Quantitative variables may be discrete and can only take on a finite (or countably infinite) set of values, such as the number of siblings and individual has, or the number of correct answers on a multiple-choice test; or quantitative variables may be continuous and can take on any value, or be measured to an arbitrary precision, within a given (or possibly infinite) range, such as the volume of reagent added to a well.

Qualitative variables, or categorical variables, are non-numeric and the data exist in categories; examples include eye colour, education level or the survival status (dead or alive) of an animal at the end of the study.

Variables can often be collected in either a quantitative or qualitative fashion. In fact, for many qualitative variables there will be an underlying quantitative response that may be recorded instead. For example, education level could be recorded as the highest degree (qualitative) or the number of years spent in school (quantitative).

A.3 Notation

In statistics, upper case letters are often used to represent variables and lower case letters to represent their realised values. For example, the variable X may represent the height of an individual while the variable Y may represent their weight. The individual observations of a data set are then often indexed (or labelled) with a subscript, with the ith observation of the variable X denoted as x_i. The sample size is typically denoted as n, and so the observations of a dataset consisting of n observations can be labelled x_1, x_2, ..., x_i, ..., x_n.

Table A.1 contains the income for a random sample of $n = 20$ individuals. Here we can use the index $i = 1, \ldots, 20$ to label each individual. If we let the variable X represent income, the income of the first individual ($i = 1$) is $x_1 = \$23,000$.

TABLE A.1
Annual employment income in dollars for a random sample of 20 individuals.

23,000	26,000	19,000	34,000	27,000	46,000	68,000	25,000	25,000	56,000
30,000	24,000	25,000	22,000	45,000	48,000	36,000	38,000	21,000	125,000

A.4   Distributions

A distribution describes the possible values that are a variable takes on, and the frequency with which each value occurs. For now, we will consider the distribution of a sample of data. However, we will later discuss the concept of a population distribution. Just as the sample mean can be used to estimate the population mean, the sample distribution can be used to learn about the population distribution.

For categorical or discrete numeric variables, the distribution can be represented by a list of the of the possible values of the variable and the (relative) frequency of each of those values. When evaluating the distribution of a continuous numeric variable, the variable is often categorised into bins, or ranges, with the (relative) frequency of observations that fall into each range tabulated.

The distribution of a data set is often depicted using plots. For categorical variables, bar charts and pie charts are common. For quantitative variables, histograms are the most common. Figure A.1 shows the histogram for the income data presented in Table A.1. The first bin (or range) is from [15,000, 25,000); i.e., from \$15,000 (inclusive) to just below \$25,000, and there are 8 observations that fall within this range.

A.4.1   Describing the distribution of numeric variables

When examining the distribution of a numeric variable, it is often of interest to identify typical values, the extent of the spread of the data, the shape of the distribution, and determine the presence of any outliers.

Distributions are often classified as being symmetric, where the right half is a mirror image of the left half, or skewed where the data are asymmetric. If a distribution has a long right tail, then the distribution is said to be right (or positively) skewed. If the distribution has a long-left tail, then it is said to be left (or negatively) skewed.

The modality of the distribution looks at the number of major peaks. If there is one major peak the distribution is said to be unimodal, if there are two major peaks the distribution is said to be bimodal, and if there are three or more major peaks the distribution is said to be multimodal.

Figure A.2 shows example histograms for bimodal, left skewed, right skewed and symmetric data. When interpreting if a distribution is symmetric, it is important to look at the general shape as the left half of the distribution will never be a perfect mirror image of the

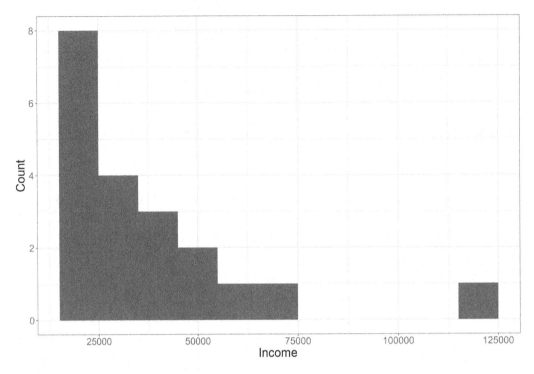

FIGURE A.1
Histogram of the income data in dollars presented in Table A.1.

right half. For the income data presented in Figure A.1, the data is right skewed with one extreme observation.

A.5   Descriptive statistics

Descriptive statistics seek to summarise or reduce a large data set to allow for information to be easily extracted. For a categorical variable we will often be interested in characterising the proportion at each level, whereas, for quantitative variables, we are often interested in the measurement of central tendency or a most typical value, or the middle point. Two common measures of central tendency are the mean (or arithmetic average) and the median.

A.5.1   Sample mean

The sample mean, often denoted as $\bar{x}$, can be thought of as the balancing point of the distribution of observations and is an estimate of the population mean. It is calculated by summing across (adding up) all observations and dividing by the sample size. Mathematically the formula for the sample mean can be represented using summation notation

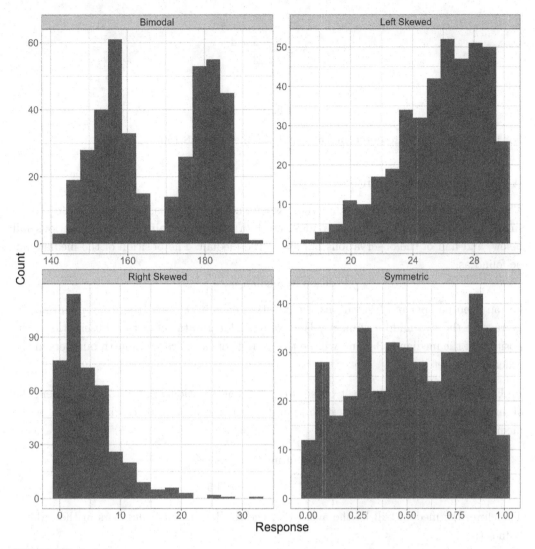

FIGURE A.2
Histogram for bimodal, right skewed, left skewed and symmetric data.

as:

$$\bar{x} = \frac{1}{n} \sum_{i=1}^{n} x_i$$
$$= \frac{x_1 + x_2 + \ldots + x_n}{n}.$$

Example: Suppose that a data set consists of $n = 3$ observations: 8, 3, 4. Then $x_1 = 8$, $x_2 = 3$ and $x_3 = 4$ and the sample mean can be calculated as:

$$\bar{x} = \frac{8 + 3 + 4}{3}$$
$$= \frac{15}{3}$$
$$= 5.$$

For the income data presented in Table A.1 the mean is \$38,150.

A.5.2 Sample median

The sample median is the 'middle' value when the observations have been ordered from smallest to largest. The median divides the data such that half of the observations fall below, and half of the observations fall above the median. The position of the median can be found as:

$$\text{Position of median} = \frac{n + 1}{2}.$$

For an even number of observations, the position ends in 0.5, and the median will be the average of the two middle numbers. For example, for a data set with 6 observations, the position of the median is 3.5 and will be the average of the third and fourth numbers in the ordered data set.

Example: Suppose a data set consists of $n = 6$ observations: 2, 6, 3, 7, 3, 5. Then, the observations ordered from smallest to largest will be: 2, 3, 3, 5, 6, 7. The position of the median will be:

$$\text{Position of median} = \frac{6 + 1}{2}$$
$$= 3.5.$$

Therefore, the median will be the average of the third and fourth numbers in the ordered data set:

$$\text{Median} = \frac{3 + 5}{2}$$
$$= 4.$$

For the income data presented in Table A.1 the median is \$28,500.

A.5.3 Mean versus median

For an (approximately) symmetric distribution, the mean can be expected to be similar to the median. However, when the distribution is skewed the mean will be 'pulled' into the long tail. For example, for a right skewed distribution the mean will be pulled towards the large observations and will be larger than the median. This may be the case when considering summarising data on salaries – a few high salaries will pull the mean towards the right tail. The mean is also sensitive to the presence of outliers, whereas the median is resistant. Consequently, for skewed distributions or if there are many outliers present, the median is often the preferred measure of central tendency.

Example: consider a data set with 5 observations: 5, 7, 8, 10, 10. For these data, the mean and the median are both 8. However, if the last observation was instead 100 (rather than 10), then the mean would be 26 while the median would be unchanged.

For the income data presented in Table A.1 the mean is \$38,150 and is somewhat larger than the median of \$28,500. Only six of the twenty observations were larger than the mean. In this case, because the data are somewhat right skewed, and there is an extreme observation, the median is a better representation of a typical value.

A.5.4 Sample proportion

The sample proportion provides the relative frequency of observations for a particular class (or category) of a qualitative variable. A particularly important class of categorical variables are binary variables, that only have two categories. For binary variables the observations can be encoded as 0 or 1. For example, consider flipping a coin and let x_i be the response for the ith flip that takes on a value of 1 if the coin shows heads and 0 if it shows tails. That is:

$$x_i = \begin{cases} 0 & \text{tails} \\ 1 & \text{heads.} \end{cases}$$

If we have n observations and x of those observations fall in category 1, then the sample proportion in category 1 is:

$$\text{Sample proportion in category 1} = \frac{x}{n},$$

and the sample proportion in category 2 is:

$$\text{Sample proportion in category 2} = 1 - \frac{x}{n} = \frac{n-x}{n}.$$

For the sample data: {tails, heads, heads, tails} this can be encoded as {0, 1, 1, 0} and the sample proportion can be computed as:

$$\begin{aligned} \text{Sample proportion heads} &= \frac{x_1 + \cdots + x_n}{n} \\ &= \frac{0+1+1+0}{4} \\ &= \frac{2}{4} \\ &= \frac{x}{n}. \end{aligned}$$

Therefore, we can also think of the sample proportion as the sample mean when the variable has been encoded as 0 or 1.

A.6 Measures of spread

Measuring the centre only gives partial information about the distribution. It is also important to consider the variability of data. For example, two assays can produce a very similar

TABLE A.2

Data sets for three samples all with the same mean and median (20) but with different spread or variation.

Sample A	18	19	20	21	22
Sample B	14	18	20	22	26
Sample C	14	17	20	23	26

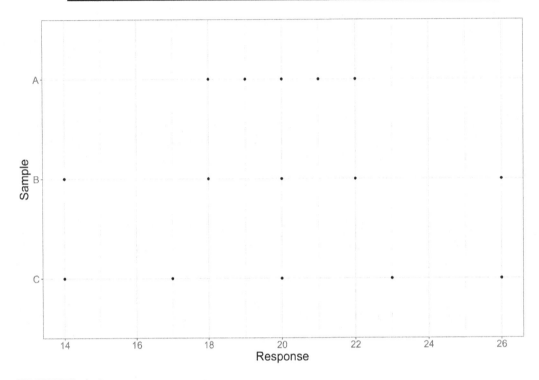

FIGURE A.3

Plot of data for three sample presented in Table A.2. All three samples have the same mean and median (20) but clearly different variability.

typical (average) results but differ substantially in terms of their variability. In this context, an assay with a small variability would be preferred.

Consider the following three samples presented in Table A.2, all three samples have the same mean and median (20) but clearly a different spread or variation. Figure A.3 provides a visual representation of these data.

A.6.1   Range

The simplest measure of variation is the range and is defined as the distance between the largest and smallest observations:

$$\text{Range} = \text{maximum} - \text{minimum}.$$

The range only accounts for the smallest and largest observation and for the examples in Table A.2 the range for Samples B and C is the same:

$$\text{Range} = 26 - 14$$
$$= 12,$$

despite the variability of Sample C clearly being greater.

A.6.2 Variance and standard deviation

The standard deviation is a measure of spread based on deviations (distances) from the sample mean $(x_i - \overline{x})$. A natural measure of variability would be the average distance (deviation) of an observation from the mean. However, the sum of the deviations across all observations will be 0; and so, the average deviation from the mean will also be 0. Therefore, when computing a measure of variability based on the deviations, we square the deviations first.

The standard deviation is calculated by first computing the sample variance as:

$$s^2 = \frac{1}{n-1} \sum_{i=1}^{n} (x_i - \overline{x})^2 \,.$$

The sample standard deviation is then found as the square-root of the variance:

$$s = \sqrt{s^2}$$
$$= \sqrt{\frac{1}{n-1} \sum_{i=1}^{n} (x_i - \overline{x})^2}.$$

Table A.3 details the intermediaries required for the calculation of the sample variance and standard deviation for the Sample A data presented in Table A.2. From this the sample variance is:

$$s^2 = \frac{1}{5-1}(10) = 2.5.$$

The sample standard deviation is:

$$s = \sqrt{2.5} = 1.581$$

The standard deviation is a positive value and will only be 0 if all the observations are the same (i.e., there is no variability in the data.) The standard deviation will increase with the amount of variation in the data and can be roughly interpreted as a 'typical' deviation of an observation from the mean. The sample standard deviation is not resistant to the presence of outliers and is the appropriate measure of variability when the mean is chosen as the measure of centre.

TABLE A.3

Calculation of intermediaries required for the calculation of the sample variance and standard deviation for the Sample A data presented in Table A.2. The sample mean for these data is $\bar{x} = 20$.

Observation x_i	Deviation from the mean $x_i - \bar{x}$	Squared deviation $(x_i - \bar{x})^2$
18	$(18 - 20) = -2$	$(-2)^2 = 4$
19	$(19 - 20) = -1$	$(-1)^2 = 1$
20	$(20 - 20) = 0$	$(0)^2 = 0$
21	$(21 - 20) = 1$	$(1)^2 = 1$
22	$(22 - 20) = 2$	$(2)^2 = 4$
Sum	$\sum_{i=1}^{n}(x_i - \bar{x}) = 0$	$\sum_{i=1}^{n}(x_i - \bar{x})^2 = 10$

A.6.3 Coefficient of variation

The coefficient of variation (CV) is the ratio of the standard deviation to the mean and normalises the standard deviation by the mean. For sample data it is found as:

$$\mathrm{CV} = \frac{s}{\bar{x}}.$$

As such, it provides the standard deviation as a percentage of the mean and can be used to make comparisons when measurements are on different scales. However, the CV has many potential pitfalls. It should only be used for data that are positive. The CV will tend to infinity as the sample mean tends to zero and will be undefined when the mean is zero.

A.6.4 Percentiles

The pth percentile divides the bottom $p\%$ of the data from the top $(100 - p)\%$. The median is the 50th percentile as half of the data falls below and half of the data falls above the median. Two other important percentiles are the first quartile or 25th percentile, denoted as Q_1, and the third quartile, Q_3, or 75th percentile. Together with the minimum and maximum, the median and first and third quartiles form what is known as the five number summary.

There are many different methods (formulae) for calculating percentiles – each method may give a slightly different result. The quartiles can be found as the medians of the first and second halves of the data. That is, Q_1 can be found as the median of the bottom half of the (ordered) data set, while Q_3 can be found as the median of the top half of the (ordered) data set. If the number of observations is odd, then the median can be included in both halves or the median can be excluded, depending on which method is being followed.

The interquartile range (IQR) is difference between Q_1 and Q_3 and is a measure of spread that is resistant to the presence of outliers or highly skewed data. The IQR can be used to identify any observations that may be potential outliers, with potential outliers being any observations more than $1.5 \times$ IQR from closest quartile.

The five number summary can be used to produce what is known as a boxplot. The bottom of the box is given by Q_1, the mid-line of the box by the median and the top of

TABLE A.4

Five number summary (along with the mean) for the income data presented in Table A.1.

Minimum	Q_1	Median	Mean	Q_3	Maximum
19,000	24,750	28,500	38,150	45,250	125,000

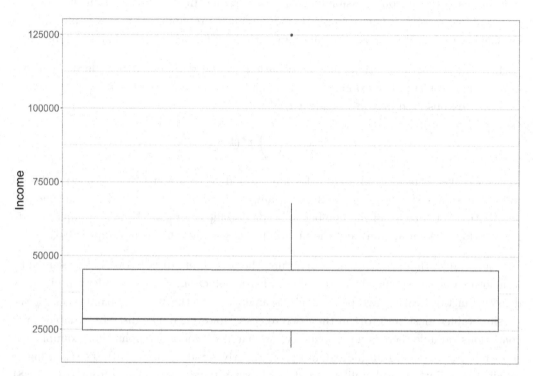

FIGURE A.4

Boxplot for the income data presented in Table A.1.

the box by Q_3. Consequently, the height of the box represents the IQR, and half of the observations fall inside the box. The whiskers extend on either side to the smallest and largest observations that are not outliers (i.e., are less than $1.5 \times$ IQR away from their closest quartile). Finally, any observations that are outliers, as defined above, are plotted as a point.

Table A.4 displays the five number summary (along with the mean) for the income data presented in Table A.1. The quartiles were calculated using the default methods in R [49] and so differ than if the methods described above were used. Figure A.4 shows the boxplot for the income data. There is one outlier, corresponding to the maximum observation.

A.7   Probability distribution

A probability distribution tells us how the total probability (of 1) is distributed among (allocated to) the various possible values of a variable. In its simplest form it can be a listing of the possible values and their corresponding probabilities. For example, when a six-sided die is rolled the possible values are 1 through 6, each with probability 1/6.

In other scenarios, a probability distribution can be represented by a formula to obtain the probabilities. For example, the binomial distribution gives the probability that x successes are observed in n independent trials:

$$P(X = x) = \binom{n}{x} p^x (1 - p)^{n-x},$$

where p is the probability of success for any individual trial and $\binom{n}{x}$ gives the number of ways to obtain x success in n trials. An example of variable that would have a binomial distribution is the count of the number of heads in 5 flips of a coin. In this example, $n = 5$, $p = 0.5$ (for a balanced coin) and the probability of getting 3 heads in 5 flips is 0.3125.

For variables that are continuous in nature, the probability distribution is represented by a smooth curve that is called the probability density function. The area under a probability density function will always be 1 – this is analogous to the total probability of all the possible values must sum to 1 in the discrete case. However, unlike discrete variables for a continuous variable there is no probability associated with any one point. For example, the probability of observing an individual that is exactly 1.8000 meters (with 0's repeating to infinity) is 0. Rather, probabilities are defined over a range, for example from 1.80 to 1.81 meters with the probability being found by computing the area under the curve. We will look at one such distribution, the normal distribution, in more detail in the next section.

A.8   Normal distribution

The normal distribution is perhaps the most important distribution in statistics and has the familiar symmetric, bell-shaped curve. Many variables have (at least approximately) a normal distribution including birth weight, and standardised test scores. The bell shaped nature of the distribution implies that most observations will be close to the centre, and that observations will be less and less likely at values further and further from the mean. The normal distribution has two parameters, the mean (μ) and the standard deviation (σ), that control the centre and spread of the distribution, respectively. The probability density function of normal curve is given by:

$$f(x) = \frac{1}{\sigma\sqrt{2\pi}} e^{\left\{ \frac{-(x-\mu)^2}{2\sigma^2} \right\}}$$

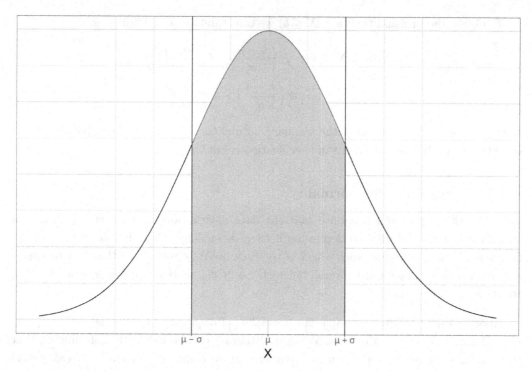

X

FIGURE A.5

The normal distribution has a bell shaped curve. Most of the area falls close to the mean μ, the area one standard deviation from the mean has been shaded and represents approximately 68% of the total area under the curve.

As for any continuous probability distribution, areas under the normal curve correspond to the percentage of observations falling within that range. We can show that the total area under the normal curve is 1. As the curve has a bell shape centred on the mean, there is more area (or probability) concentrated around the mean and progressively less area as you move further out into the tails. To obtain areas under the curve, there is no simple closed form solution, and the area must instead be found using software or a standard normal table instead.

Figure A.5 shows a plot of a normal distribution. For any normal distribution, approximately 68% of observations fall within one standard deviation of the mean between $\mu - \sigma$ and $\mu + \sigma$ (shaded in figure). Approximately 95% of observations fall within two standard deviations of the mean between $\mu - 2\sigma$ and $\mu + 2\sigma$ and approximately 99.7% of observations fall within three standard deviations of the mean between $\mu - 3\sigma$ and $\mu + 3\sigma$.

Standard normal distribution

The standard normal distribution is the normal distribution with mean $\mu = 0$ and standard deviation $\sigma = 1$. For a variable X with a normal distribution with mean μ and standard deviation σ we can obtain the standardised version as:

$$Z = \frac{X - \mu}{\sigma},$$

where Z has a standard normal distribution.

To obtain the probability $P(a \leq X \leq b)$ we can standardise the limits as:

$$P(a \leq X \leq b) = P\left(\frac{a - \mu}{\sigma} \leq Z \leq \frac{b - \mu}{\sigma}\right)$$
$$= \Phi\left(\frac{b - \mu}{\sigma}\right) - \Phi\left(\frac{a - \mu}{\sigma}\right),$$

where $\Phi(z)$ gives the area under the standard normal curve to the left of z. We use this to solve for the probability of an out-of-specification result.

A.8.1   Assessment for normality

Many statistical procedures require that the data have a normal distribution. A formal statistical test can be used to determine if there is evidence that the observed data are non-normal. An example of such a test is the Shapiro Wilk test. Note that this test does not confirm that the data are normal, but rather only claims that there is no evidence that the data are not normal.

Another useful tool for assessing the normality of data is to plot the data. One option is to plot a histogram of the data. While a histogram can be helpful in determining that the data are not normal, it is not sensitive enough to confirm the data are (reasonably) normal. Consequently, the preferred plot for assessing the normality of data is the normal quantile-quantile plot (QQ plot).

The normal QQ plot is a plot of the observed quantiles of the data against theoretical quantiles of the normal distribution. If the data have a normal distribution, then the observed quantiles will closely match the theoretical quantiles and the relationship between the two will be linear. If the data does not have a normal distribution than the observed quantiles will diverge from this line. It is important to look for general trends as the points will never fall exactly on the line.

Figure A.6(a) shows a QQ plot for data from a normal distribution. Here it can be seen that the points roughly have a linear pattern. Figure A.6(b) shows data that come from a right skewed distribution; this plot shows evidence of non-normality as the points clearly have a non-linear pattern.

Note that quantile-quantile plots are not restricted to assessing if the distribution is normal. We can use a QQ plot to assess if the observed data follows any distribution. For example, if we wanted to assess the distribution of the data in Figure A.6(b) as an exponential, then we can create an exponential QQ plot.

A.9   Confidence intervals

A confidence interval (CI) gives a range of plausible values for a parameter. For example, if we had a 95% confidence interval for the relative potency from (0.88, 1.14), we may not

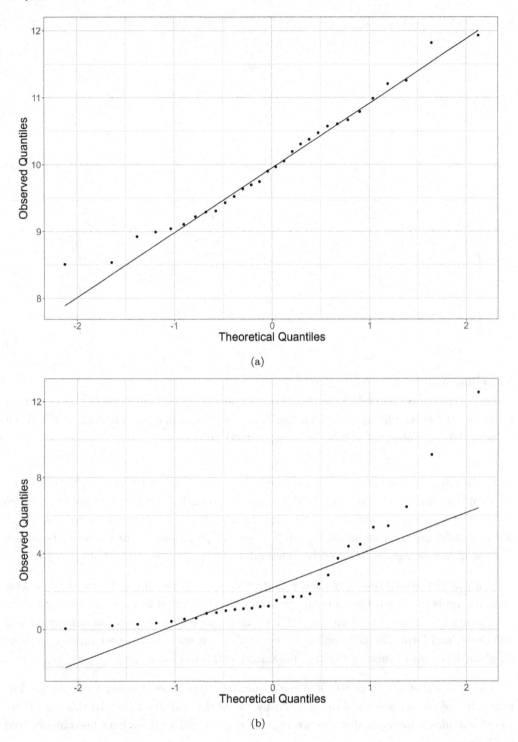

(a)

(b)

FIGURE A.6

Normal quantile-quantile plot for normally distributed data in (a) and non-normally distributed data in (b).

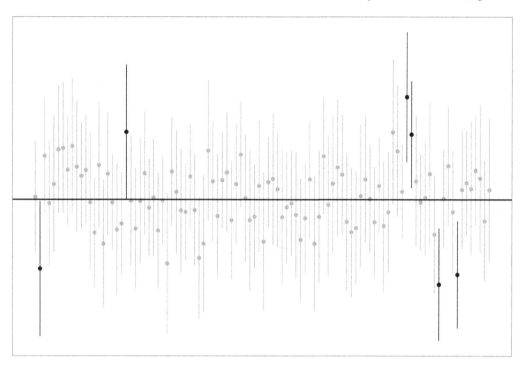

FIGURE A.7

100 95% confidence intervals for estimating a parameter. The true value of the parameter is represented by the horizontal black line. There were 6 confidence intervals that did not capture the true value; these have been highlighted in black.

know the true relative potency, but we have a high degree of confidence (95%) that it lies between the limits of our interval. In this sense, the confidence interval gives both an idea of where the parameter may lie and the uncertainty with which the estimation is made. A narrow confidence interval indicates a high degree of precision, whereas a wide confidence interval indicates there is considerable uncertainty.

The confidence level represents how confident we are that we captured the true parameter (e.g., μ) and is the proportion of confidence intervals that would be expected to capture the true parameter with repeated sampling. For example, if we were to carry out our experiment 100 times, and form 100 95% confidence intervals, then we would expect (approximately) 95 of our intervals to capture the true parameter and 5 to miss it.

Figure A.7 shows a set of 100 95% confidence intervals for estimating a parameter. The true value of the parameter is represented by the horizontal black line. In this case there were 6 confidence intervals that did not capture the true value; these have been highlighted in black.

For many parameters, including means, and regression coefficients, a general format for the confidence interval is given by:

$$\text{Estimate} \pm \text{reference value} \times \text{standard error(estimate)},$$

where the term to the right of the $\pm$ sign is called the margin of error. The reference value reflects the confidence level and will typically be a value from t-distribution. The standard error of the estimate accounts for the variability of the data and represents the variability of the estimate. Therefore, a confidence interval can also be written as:

$$\text{Estimate} \pm \text{Margin of error.}$$

A notable exception to this general format for a confidence interval is the confidence interval for the variance.

We will first demonstrate the concept of a confidence in terms of a single mean μ, where the variance is unknown based on a sample of size n. In this case, the estimate of the population mean will be given by the sample mean $\bar{x}$ and for a $(1 - \alpha) \times 100\%$ confidence interval the reference value will be the $1 - \alpha/2$ percentile of the t-distribution with $n - 1$ degrees of freedom. Finally, the standard error of the mean is given by the sample standard deviation, s, divided by the sample size, i.e.:

$$\text{Standard error}\,(\bar{x}) = \frac{s}{\sqrt{n}}.$$

Therefore, the confidence interval for μ is given by:

$$\bar{x} \pm t_{1-\alpha/2, n-1} \times \frac{s}{\sqrt{n}}.$$

From the formula above, the margin of error, and therefore the width of the confidence interval depend on: the value of the t-statistic, the standard deviation and the sample sample size.

In general, the margin of error will get smaller as the confidence level is decreased. This is because the t-statistic will be smaller. For example, for a 95% CI based on a sample of size $n = 31$, the t-statistic is: $t_{0.975,30} = 2.04$, while for a 90% CI: $t_{0.95,30} = 1.70$. If the standard deviation is smaller the confidence interval will also be narrower. However, this is typically a function of the population being studied and is not something that can be easily controlled for. However, if the sample size is increased, the standard error and the margin of error will decrease. Consequently, the width of the confidence interval will be narrower and the estimate more precise.

A.10 Hypothesis tests

Hypothesis tests are often used to make decisions about the value of population parameters. inference to make decisions about the value of parameter. For example: is the new reference standard equally potent to the old standard; does the mean weight of a chocolate bar differ from the advertised weight of 100 grams; or does a new vaccine reduce probability of infection?

A statistical hypothesis is a claim about the parameter of interest. The null hypothesis, H_0, is the claim that is initially assumed to be true and is often a hypothesis of no effect or no difference. The alternative hypothesis, H_A, is the hypothesis to be considered as an alternative to H_0 and is sometimes referred to as the research hypothesis and is often what we wish to show. In general, statistical hypotheses are always stated in terms of parameters and should be chosen before looking at the data.

The null hypothesis is often a statement that the population parameter takes on a particular value, H_0: $\mu = \mu_0$. For example, the mean weight of a chocolate bar is the advertised weight, 100 grams:

$$H_0: \mu_0 = 100g.$$

There are three choices for the alternative hypothesis. There are two one sided tests that seek to determine if the parameter is less than the null value, H_A: $\mu < \mu_0$; the parameter is greater than the null value, H_A: $\mu > \mu_0$. Finally there is the two sided test that seeks to determine if the population parameter different from the null value, H_A: $\mu \neq \mu_0$.

For the chocolate bar example, the three choices of alternative are:

1. The chocolate bars are under weight, on average:

$$H_A: \mu < 100g.$$

2. The chocolate bars are over weight, on average:

$$H_A: \mu > 100g.$$

3. The chocolate bars have, on average, a different weight than labelled:

$$H_A: \mu \neq 100g.$$

The choice for the alternative hypothesis test will depend on the research question. The first alternative hypothesis would be of interest to a consumer, while the third alternative would be of most interest to the producer.

In a hypothesis test, sample data are used to determine if:

1. The data are consistent with the null hypothesis. In this case the decision will be to fail to reject H_0; or

2. The data are inconsistent with the null hypothesis and are instead consistent with the alternative hypothesis. In this case the decision will be to reject H_0 and conclude H_A.

Note that in a hypothesis test, we do not accept H_0, but rather only fail to reject it. A classic example of hypothesis testing occurs in a court of law. Here, a defendant is assumed innocent (null hypothesis), until proven guilty (alternative hypothesis). In this context, failure to prove guilt does not imply innocence.

It is important that the decision to reject (or fail to reject) the null hypothesis is objective. A test statistic is a value calculated using sample data to make decision if H_0 should

TABLE A.5

Summary of possible outcomes of a hypothesis test.

	H_0 True	H_0 False
Reject H_0	Type I Error	Correct Decision
Do not reject H_0	Correct Decision	Type II Error

be rejected (or not). For many parameters, test statistics will take the general format of

$$\text{Test statistic} = \frac{(\text{estimate} - \text{hypothesised value})}{\text{standard error(estimate)}}.$$

For example if performing a hypothesis test for one population mean, μ: the estimate of μ is given by the sample mean $\bar{x}$; the hypothesised value will be the mean if the null hypothesis is true, μ_0; and the standard error of the mean will account for variability and is given by $\frac{s}{\sqrt{n}}$. Therefore, the test static is:

$$\text{Test statistic} = \frac{\bar{x} - \mu_0}{s/\sqrt{n}}.$$

If the null hypothesis is true, then the test statistic will have a known distribution that can be used to obtain probabilities. In the case of one mean, the test statistic will have a t-distribution with $n - 1$ degrees of freedom.

The p-value is defined as the probability of obtaining a test statistic as, or more, contradictory to the null hypothesis than that observed. Therefore, the smaller the p-value, the more evidence against null hypothesis. The significance level of a hypothesis test is the threshold for the p-value at which we decide to reject the null hypothesis. If the p-value is smaller than the significance level then we conclude that the data are unlikely to occur by chance alone and reject the null hypothesis. If the p-value is larger than the significance level, then we will fail to reject the null hypothesis. The significance level should be decided before any data are collected. Note, that the p-value is not: the probability the H_0 is true (or false), or the probability that an erroneous conclusion is reached.

In truth, the null hypothesis is either true or false. Our conclusion of the hypothesis test will be to either reject, or not to reject the null hypothesis. This leads to two possible types of errors being made in a hypothesis test. A Type I error occurs when the null hypothesis is rejected, but is, in fact, true; a Type II error occurs when the null hypothesis is not rejected, but is false. Table A.5 summarises the possible outcomes of a hypothesis test.

There is a probability associated with making Type I or Type II errors. The significance level, α, is the probability of making a Type I error, while the probability of making a Type II error is denoted as β. The power of the test, that is, the probability of detecting a true difference, can be found as 1-β. Ideally, we would like both α and β to be as small as possible. However, making it harder to reject a true null (small α) also makes it easier to fail to reject a false null (large β).

In practice, we typically will restrict the value of α and choose a test that maximises the power. Most often, we will set the significance level α to be 5% so that probability of rejecting a true null hypothesis is small. Therefore, the null hypothesis is given the benefit of the doubt and the burden of proof is on the alternative hypothesis.

Note that the results of the hypothesis test are statistically significant if the null hypothesis is rejected at the chosen level of α. This does not necessarily mean the difference has any practical implications. Large sample sizes can show statistical significance for small differences. In the context of bioassay, a very precise assay can result in a significant result even if that result has no practical implications. Therefore, an equivalence testing framework can often be preferred and is discussed in Chapter 7.

In summary the statistical hypothesis test consists of the following 6 steps:

1. State null and alternative hypotheses;

2. Decide the significance level;

3. Compute the test statistic;

4. Find the P-value;

5. Compare P-value to significance level;

6. State conclusion – interpret results of the hypothesis test.

Finally, confidence intervals are related to hypothesis tests. The limits of a confidence interval can be found by rearranging the test statistic to find the values that lie on the decision boundary for the two-sided alternative hypothesis. That is, we find the values for the estimate where the decision flips from reject to not-reject. To reiterate, a confidence interval gives a range for plausible values and so gives the range of values for which a two-sided hypothesis test will not be rejected.

A.11   One mean

When conducting a hypothesis test for one population mean where $H_0{:}\mu = \mu_0$, the test statistic is given by:

$$t = \frac{\bar{x} - \mu_0}{s/\sqrt{n}}.$$

We have three choices of alternative:

- $H_A{:}\ \mu < \mu_0$;

- $H_A{:}\ \mu > \mu_0$;

- $H_A{:}\ \mu \neq \mu_0$.

The confidence interval for μ is given by:

$$\bar{x} \pm t_{(1-\alpha/2),n-1}\frac{s}{\sqrt{n}}.$$

The degrees of freedom for the test statistic and t-value for CI are $df = n - 1$.

Hypothesis tests and confidence intervals for one mean, based on the t-distribution assume that the observations are independent and identically distributed. These procedures also assume that the observations have been sampled from a normal population. However, the procedures are robust to the assumption of normality, particularly if the sample size is large ($n > 30$).

A.12 Comparison of two means

A common research question is to make a comparison between the parameters of two different populations, for example: is the drug substance protein content the same before and after a change to manufacturing methods; or does the new drug result in an increase to the amount of weight lost when compared to a placebo?

Initially, we will assume that the two samples are independent of each other. That is, the observations in one sample are not related to (give no information about) the observations in the other. We will then look at methods for paired data. As for the case of one mean, we will assume that the two populations both have a normal distribution. However, provided the sample sizes are large (both greater than 30) the procedures are relatively robust to this assumption.

Let n_1 be the sample size for the random sample from population 1 and n_2 the sample size for the random sample from the population 2. Let μ_1 and μ_2 be the population means that we wish to compare, with estimates (sample means) $\bar{x}_1$ and $\bar{x}_2$. Finally, let s_1 and s_2 be the sample standard deviations of the two samples.

A.12.1 Unequal variance

We first present the case where we assume that the populations have unequal variances. That is, the variance of population 1 is different from than the variance of population 2: $\sigma_1^2 \neq \sigma_2^2$. In this case, each variance must be estimated separately using the corresponding sample standard deviation.

The test statistic for comparing two means where $H_0 : \mu_1 - \mu_2 = \Delta_0$ is:

$$t = \frac{(\bar{x}_1 - \bar{x}_2) - \Delta_0}{\sqrt{\frac{s_1^2}{n_1} + \frac{s_2^2}{n_2}}}.$$

We again have three choices of alternative:

- H_A: $\mu_1 - \mu_2 < \Delta_0$;

- H_A: $\mu_1 - \mu_2 > \Delta_0$;

- H_A: $\mu_1 - \mu_2 \neq \Delta_0$.

When comparing two means, we are often interested in testing if the two means are the same, i.e., $\mu_1 = \mu_2$. This is equivalent to testing if $\Delta_0 = 0$.

The confidence interval for difference between means $\mu_1 - \mu_2$:

$$(\bar{x}_1 - \bar{x}_2) \pm t_{1-\alpha/2, df} \sqrt{\frac{s_1^2}{n_1} + \frac{s_2^2}{n_2}}$$

The degrees of freedom for the test statistic and the t-value in the confidence interval can be approximated using the Satterthwaite approach [51].

A.12.2   Equal variances

When comparing two population means, we can also make the assumption that the population variances are equal, that is, $\sigma_1^2 = \sigma_2^2$. In this case, the estimate of the variance can be obtained by pooling the information across both samples. The pooled standard deviation can be found as:

$$s_p = \sqrt{\frac{(n_1 - 1)s_1^2 + (n_2 - 1)s_2^2}{n_1 + n_2 - 2}}.$$

To test the hypothesis H_0: $\mu_1 - \mu_2 = \Delta_0$

$$t = \frac{(\bar{x}_1 - \bar{x}_2) - \Delta_0}{s_p \sqrt{\frac{1}{n_1} + \frac{1}{n_2}}}.$$

The confidence interval for $\mu_1 - \mu_2$ is

$$(\bar{x}_1 - \bar{x}_2) \pm t_{1-\alpha/2, n_1+n_2-2} s_p \sqrt{\frac{1}{n_1} + \frac{1}{n_2}}.$$

In this case the degrees of freedom for the test statistic and confidence interval are: $df = n_1 + n_2 - 2$.

The ratio of the sample standard deviations can be used to determine if the pooled variance procedure is appropriate. A general rule of thumb that can be applied when the sample sizes are similar is that the pooled procedure is appropriate so long as neither standard deviation is more than double the other

$$0.5 < \frac{s_1}{s_2} < 2.$$

More formally an F-test can be conducted to test if the variances are the same.

A.12.3   Paired data

If the observations are paired as $(X_1, Y_1), (X_2, Y_2), \ldots, (X_n, Y_n)$, for example, if we were to measure response for each individual before and after treatment (X = before and Y = after), then we require an alternative procedure for comparing the means. In this case, for each observation, we can compute the difference, $D_i = X_i - Y_i$, and apply the one-sample t-test and one sample confidence interval procedures using the differences. In this case, we assume the differences $D_i, i = 1, \ldots, n$ are normally distributed with mean $\Delta = \mu_D = \mu_1 - \mu_2$ and variance σ^2.

Let the sample mean difference be calculated as:

$$\bar{d} = \sum_{i=1}^{n} d_i,$$

with sample standard deviation:

$$s_d = \frac{1}{n-1} \sum_{i=1}^{n} \left(d_i - \bar{d}\right)^2.$$

Then, the test statistic for testing the null hypothesis: H_0: $\mu_D = \Delta_0$ is given by:

$$t = \frac{\bar{d} - \Delta_0}{s_d/\sqrt{n}}.$$

We again have three choices of alternative:

- H_A:$\Delta < \Delta_0$;

- H_A:$\Delta > \Delta_0$;

- H_A:$\Delta \neq \Delta_0$.

The confidence interval is given by:

$$\bar{d} \pm t_{(1-\alpha/2),n-1} \frac{s_d}{\sqrt{n}}.$$

The degrees of freedom for the test statistic and t-value for CI are $df = n - 1$.

When working with paired data, we can show that the variance of the sample mean difference is:

$$Var(\bar{d}) = \frac{\sigma_1^2 + \sigma_2^2 - 2\rho\sigma_1\sigma_2}{n},$$

where ρ is the correlation between the paired measurements. For independent samples $\rho = 0$, but typically $\rho > 0$ for most paired experiments. Therefore, accounting for the paired nature of the data will result in the standard error of the test statistic and confidence interval being smaller than if independence is assumed. Consequently, we will be more likely to achieve a statistically significant result, and the confidence interval will be narrower.

A.13   Analysis of variance (ANOVA)

Analysis of variance (ANOVA) generalises the (pooled variance) t-test to 2 or more populations. Let k be the number of population means we wish to compare. Then, the null hypothesis for an ANOVA is that all population means are the same:

$$H_0: \mu_1 = \mu_2 = \cdots = \mu_k,$$

and the alternative hypothesis, is that at least one group mean is different:

$$H_A: \mu_i \neq \mu_j \text{ for some } i \neq j.$$

Note that an ANOVA does not tell you what group mean(s) is/(are) different. Rather a post-hoc analysis, for example, Tukey's HSD (honestly significant difference), will need to be used to determine this.

ANOVA compares the variability between and within groups. A significant difference will be identified if the variability between groups is large relative to the variability within groups. The assumptions for ANOVA are that the populations are normal and that the variance are constant across groups. ANOVA can also be used to estimate variance components, and we use this in Chapters 3 and 8.

A.14   Regression

Regression is used when we are interested in studying the relationship between two (or more) variables. That is, how one variable changes in relation to another; or the prediction of the value of one variable using the other(s). Examples include investigating the relationship between high school grades and university grades, and the concentration (dose) and response. We are interested in a model, where there is uncertainty in the value of the response Y for a fixed value of $X = x$. Suppose Y is a (random) variable representing the response measured for an assay, and we are interested in the response at a dose of $X = 0.125$. In general, there will be some uncertainty (variability) in the response for repeated measurements at a dose of $X = 0.125$. Scatterplots can be useful to help us visualise the relationship between two (quantitative) variables. Each point will represent the (x, y) values for an individual. Typically we plot the predictor variable x on the horizontal axis and the response variable y on the vertical axis.

A.14.1   Simple linear regression

In the simple linear regression model, the mean value of Y (response) is modelled as a linear function of x (predictor):

$$\mu_Y = \beta_0 + \beta_1 \times x.$$

The simple linear regression model can be defined as:

$$Y = \beta_0 + \beta_1 \times x + \epsilon$$

where ϵ represents a random error and has a normal distribution with a mean $\mu = 0$ and variance σ^2. We define the estimated regression line:

$$\hat{y} = \hat{\beta}_0 + \hat{\beta}_1 \times x,$$

where $\hat{y}$ is the predicted (fitted) value of Y for a given value of x, $\hat{\beta}_0$ is an estimate of the intercept β_0 and $\hat{\beta}_1$ is an estimate of the slope β_1. We can define a residual (realisation of the error component) as the difference between the observed and predicted value

$$e_i = y_i - \hat{y}_i.$$

The estimate of the intercept is the predicted value of the response when the predictor variable is 0. It is important to take care when interpreting the intercept as it can often be an extrapolation. The estimate of the slope represents the average change in the response for a 1 unit increase in the predictor.

To find the best fitting straight line, we minimise the sum of the squared residuals; this is known as the method of least squares:

$$\sum_{i=1}^{n}(y_i - \hat{y}_i)^2 = \left[y_i - (\hat{\beta}_0 + \hat{\beta}_1 \times x)\right]^2.$$

For the simple linear regression case, there are closed form solutions:

$$\hat{\beta}_0 = \bar{y} - \hat{\beta}_1 \bar{x}$$

and

$$\hat{\beta}_1 = \frac{\sum_{i=1}^{n}(x_i - \bar{x})(y_i - \bar{y})}{\sum_{i=1}^{n}(x_i - \bar{x})^2}.$$

The sum of squared errors, SSE, (or the sum of the squared residuals) is the amount of variation in Y that is unexplained by the model.

We are often interested in inference about the slope parameter; that is we wish to test if there there a linear relationship and $\beta_1 \neq 0$. If we wish to test the hypotheses:

$$H_0\colon \beta_1 = 0$$

$$H_A\colon \beta_1 \neq 0,$$

then the test statistic is given by:

$$t = \frac{\hat{\beta}_1 - 0}{s_{\hat{\beta}_1}}$$

where $s_{\hat{\beta}_1}$ is the standard error of $\hat{\beta}_1$ and is typically provided by software. A confidence interval for β_1 is given by:

$$\hat{\beta}_1 \pm t_{1-\alpha/2,n-2} \times s_{\hat{\beta}_1}$$

A.14.2 Non-linear regression

Linear regression refers to the fact that the model is linear with respect to the parameters, i.e., can be written as:

$$Y = \beta_0 + \beta_1 x_1 + \beta_2 x_2 + \ldots + \beta_p x_p + \epsilon.$$

Non-linear regression refers to the response being a non-linear function of the parameters:

$$Y = f(x) + \epsilon.$$

An example is the four parameter logistic model:

$$Y = D + \frac{A - D}{1 + e^{B(\log(\text{dose}_i) - C)}} + \epsilon.$$

In this case, we cannot rearrange the equation to be a linear function of the parameters. Model fitting for non-linear regression models is more complicated and must be done using successive approximations (there is no closed form solution for the parameter estimates). Specialist statistical software is available for this purpose.

A.15   Logarithms

The logarithm of a number, x, with base b, is defined as the exponent to which b must be raised to yield x.

For example, the logarithm off 100 with base 10 is 2; this is because $10^2 = 100$. In other words, we must raise the base 10 to the power of 2 to yield the number x = 100.

In mathematical terms $\log_{10} 100 = 2$. Common choices for the base include 10, the natural logarithm $e \approx 2.718$, as well as 2.

Addition and subtraction on the log scale correspond to multiplication and division on the original scale. For example, if considering a dilution series with top dose of 1024 that is serially diluted by a factor of 2, it may make sense to consider log base 2, where $\log_2 1024 = 10$, the next dose of $\log_2 512 = 9$ and so on. That is, the logarithm of the dose is the power to which 2 must be raised to yield the dose. We can think of a 1 unit increase (or decrease) on the log (base 2) scale as representing a doubling (or halving).

Taking the logarithm is a commonly used method to ensure that the normality assumption is met. A value that is skewed on the original scale may have a normal distribution on the log scale. This is true for the case of relative potencies, as well as the EC_{50}.

When the mean is calculated on the log scale and back transformed to the original scale by taking the antilog, this is equivalent to calculating the geometric mean. The geometric mean is used to calculate the average when working with data that are multiplicative in nature.

B

Example data sets

DOI: 10.1201/9781003449195-B

TABLE B.1

Data for illustration of non-constant variance analysed in Section 5.1.1.

Dose	Response							
8	3.72	2.84	2.78	2.02	3.28	1.77	5.48	1.99
16	7.57	3.05	2.18	1.93	3.66	2.24	3.94	3.63
32	10.66	5.56	12.89	10.49	9.51	10.83	5.54	6.32
64	19.70	22.87	30.00	24.77	11.70	26.74	16.85	30.64
128	40.53	22.17	20.38	37.77	47.03	14.91	61.39	48.45
256	44.89	52.16	40.12	34.26	42.74	29.61	84.23	32.41

TABLE B.2

Data for illustration of a non-normal response analysed in Section 5.1.2.

Dose	Response							
8	1.05	0.98	1.21	0.45	0.37	1.49	1.05	1.23
16	1.19	1.23	1.15	1.08	1.33	0.60	0.10	1.20
32	1.60	1.28	1.14	1.32	1.20	1.56	1.38	1.49
64	1.99	1.94	1.79	2.14	1.71	1.98	1.63	1.90
128	2.55	1.95	2.73	2.54	2.60	2.36	2.36	2.09
256	2.59	2.20	2.52	2.16	2.29	2.26	2.00	2.49

TABLE B.3

Data for illustration of a straight-line dose response relationship analysed in Section 6.1.1.

Dose	Response			
	Reference standard		Test sample	
4	1.22	1.30	1.14	1.21
16	1.51	1.59	1.50	1.33
64	1.90	1.88	1.97	1.76
256	2.24	2.24	2.00	1.96
1024	2.36	2.50	2.45	2.39
4096	2.87	2.77	2.75	2.82
16384	2.98	3.15	3.12	3.03
65536	3.45	3.51	3.37	3.36

TABLE B.4

Data for illustration of a 4PL dose response relationship analysed in Section 6.1.2.

Dose	Response					
	Reference standard			Test sample		
2	1.04	0.96	1.01	0.95	0.98	0.92
4	1.09	1.11	1.00	1.01	1.02	1.02
8	1.19	1.14	1.16	1.08	1.03	1.03
16	1.39	1.40	1.45	1.16	1.21	1.28
32	1.88	2.01	1.96	1.66	1.61	1.64
64	2.63	2.68	2.60	2.30	2.34	2.36
128	3.37	3.28	3.34	3.09	3.02	3.11
256	3.74	3.71	3.75	3.56	3.50	3.48
512	3.87	3.93	3.87	3.77	3.81	3.72
1024	3.95	4.01	3.91	3.92	3.87	3.87

TABLE B.5

Data for illustration of a 5PL dose response relationship analysed in Section 6.1.3.

Dose	Response					
	Reference standard			Test sample		
30000	149.64	133.49	143.11	156.11	145.66	155.29
10000	147.50	143.27	143.94	147.92	160.65	154.80
3333.33	152.90	145.30	154.92	167.07	160.73	153.89
1111.11	198.72	188.72	178.89	199.19	208.57	205.17
370.37	261.49	266.39	270.20	308.42	304.68	292.71
123.46	372.71	371.59	363.56	398.13	388.13	386.48
41.15	433.51	435.99	425.09	440.54	438.20	436.64
13.72	460.42	452.15	452.14	463.26	465.57	480.46
4.57	475.12	468.67	472.07	478.67	490.31	478.95
1.52	480.47	474.98	470.80	492.89	481.24	493.56

TABLE B.6

Data for illustration of an interpolation bioassay analysed in Section 6.1.4 and Section 6.1.4.1.

Standard		Test sample	
Concentration	Response	Dilution	Response
20000.00	9.46	1:1	8.5
10000.00	9.36	1:2	6.4
5000.00	9.11	1:4	4.3
2500.00	8.12		
1250.00	5.97		
625.00	3.90		
312.50	2.94		
156.25	2.53		
78.12	2.60		
39.06	2.52		

TABLE B.7

Data for illustration of a bioassay with a binary response analysed in Section 6.2.

	Reference standard		Test sample	
Dose	Dead	Alive	Dead	Alive
0.05	1	39	3	37
0.14	7	33	10	30
0.22	9	31	14	26
0.61	28	12	35	5
1.65	40	0	39	1

TABLE B.8

Data for illustration of a straight-line dose response relationship with different intercepts for the standard and the same slope for each plate; analysed in Section 6.4.

	Plate			
Dose	1	2	3	4
Reference standard				
1	1.0	3.3	0.6	3.0
2	3.5	5.3	2.8	5.3
4	5.2	7.2	4.7	6.3
8	7.5	9.0	6.6	9.4
Test sample				
1	0.3	1.9	0.0	1.9
2	1.9	4.6	2.1	3.8
4	4.7	6.6	4.1	5.8
8	6.6	8.2	5.8	8.3

TABLE B.9

Data for illustration of a straight-line dose response relationship with different slopes and intercepts for each plate; analysed in Section 6.4.

	Plate			
Dose	1	2	3	4
Reference standard				
1	29.4	31.9	14.0	41.9
4	63.6	90.4	49.1	76.9
16	109.3	146.0	81.2	108.6
64	159.3	204.1	134.9	147.4
Test sample				
1	20.2	10.5	10.1	39.2
4	60.8	72.1	52.7	82.6
16	123.0	132.3	93.1	102.1
64	163.2	201.0	136.3	138.3

TABLE B.10

Data for illustration of outlier methods analysed in Section 6.5.

Dose	Response		
2	10731.13	10398.38	10610.30
4	11146.16	9996.89	4443.63
8	14529.48	12781.58	15004.97
16	23260.28	21715.27	17741.51
32	38326.54	37293.96	69000.22
64	59182.72	60279.49	75000.00
128	79709.77	76655.11	78630.18
256	89626.65	90321.77	89942.34
512	97047.17	94135.19	96546.88
1024	101472.53	99862.92	92000.82

TABLE B.11

Estimate and confidence interval for the C parameter for 100 samples. These data are analysed in Section 7.2.1.

Sample	Estimate	CI	Sample	Estimate	CI	Sample	Estimate	CI	Sample	Estimate	CI
1	5.10	(4.99, 5.20)	2	5.04	(5.00, 5.08)	3	5.01	(4.86, 5.16)	4	5.13	(5.05, 5.21)
5	5.02	(4.95, 5.10)	6	5.06	(5.01, 5.1)	7	5.01	(4.88, 5.14)	8	5.04	(4.99, 5.09)
9	5.04	(5.00, 5.09)	10	5.05	(4.99, 5.1)	11	5.02	(4.88, 5.17)	12	5.00	(4.90, 5.09)
13	5.15	(5.01, 5.29)	14	5.05	(4.98, 5.13)	15	5.13	(4.95, 5.30)	16	5.05	(4.95, 5.15)
17	4.97	(4.83, 5.12)	18	5.12	(5.03, 5.20)	19	5.04	(4.91, 5.17)	20	5.04	(4.90, 5.19)
21	5.03	(4.97, 5.08)	22	5.08	(5.00, 5.16)	23	5.07	(4.95, 5.19)	24	5.08	(5.03, 5.13)
25	5.02	(4.92, 5.13)	26	5.10	(5.00, 5.19)	27	5.04	(4.95, 5.13)	28	5.10	(4.92, 5.27)
29	5.07	(5.00, 5.13)	30	4.98	(4.88, 5.07)	31	5.07	(4.96, 5.17)	32	5.04	(4.91, 5.17)
33	5.02	(4.84, 5.20)	34	5.07	(5.01, 5.13)	35	5.08	(5.02, 5.13)	36	5.11	(4.98, 5.23)
37	5.04	(4.97, 5.11)	38	5.09	(4.99, 5.19)	39	5.02	(4.93, 5.11)	40	5.05	(5.00, 5.10)
41	5.10	(5.02, 5.17)	42	5.03	(4.99, 5.07)	43	5.01	(4.91, 5.10)	44	5.05	(4.99, 5.11)
45	5.04	(4.89, 5.19)	46	5.05	(4.98, 5.12)	47	5.08	(5.04, 5.12)	48	5.10	(5.03, 5.16)
49	5.14	(5.02, 5.26)	50	5.07	(5.00, 5.14)	51	4.96	(4.86, 5.06)	52	5.10	(5.00, 5.21)
53	5.01	(4.94, 5.07)	54	5.08	(5.00, 5.17)	55	4.95	(4.83, 5.07)	56	5.04	(4.97, 5.10)
57	5.10	(5.04, 5.16)	58	5.10	(5.00, 5.20)	59	5.06	(4.98, 5.13)	60	5.04	(4.94, 5.14)
61	5.05	(4.99, 5.11)	62	5.05	(4.87, 5.23)	63	5.06	(4.91, 5.22)	64	4.98	(4.88, 5.09)
65	5.16	(5.06, 5.27)	66	5.06	(5.01, 5.12)	67	5.08	(5.01, 5.14)	68	5.09	(5.01, 5.18)
69	5.11	(4.98, 5.23)	70	5.04	(4.90, 5.18)	71	5.03	(4.97, 5.10)	72	5.08	(5.03, 5.13)
73	5.08	(5.00, 5.15)	74	5.02	(4.94, 5.11)	75	4.98	(4.85, 5.11)	76	5.06	(4.96, 5.16)
77	5.08	(5.03, 5.13)	78	4.99	(4.91, 5.06)	79	5.09	(5.00, 5.18)	80	5.01	(4.88, 5.14)
81	5.08	(4.94, 5.22)	82	5.00	(4.93, 5.07)	83	4.99	(4.89, 5.10)	84	5.03	(4.92, 5.13)
85	5.05	(4.99, 5.12)	86	5.01	(4.94, 5.07)	87	4.91	(4.76, 5.06)	88	5.09	(4.96, 5.21)
89	5.06	(5.01, 5.11)	90	5.09	(4.96, 5.22)	91	4.95	(4.86, 5.03)	92	5.07	(4.99, 5.14)
93	5.00	(4.91, 5.09)	94	5.16	(5.05, 5.27)	95	5.09	(5.03, 5.15)	96	5.09	(4.97, 5.21)
97	5.08	(5.04, 5.12)	98	5.05	(4.99, 5.11)	99	5.06	(4.94, 5.19)	100	5.03	(4.94, 5.13)

TABLE B.12

Estimate and confidence interval for the difference in slope for 100 samples known to be parallel and 10 samples known to be non-parallel. These data are analysed in Section 7.2.2.

Sample	Estimate	CI	Sample	Estimate	CI	Sample	Estimate	CI	Sample	Estimate	CI
Known to be parallel											
1	−0.02	(−0.05, 0.00)	2	−0.01	(−0.05, 0.04)	3	−0.02	(−0.06, 0.02)	4	0.01	(−0.05, 0.06)
5	0.01	(−0.02, 0.04)	6	−0.01	(−0.06, 0.04)	7	−0.01	(−0.06, 0.04)	8	0.00	(−0.06, 0.06)
9	0.01	(−0.01, 0.03)	10	0.01	(−0.05, 0.08)	11	−0.01	(−0.03, 0.02)	12	−0.03	(−0.10, 0.05)
13	0.07	(0.01, 0.13)	14	−0.04	(−0.11, 0.04)	15	0.04	(−0.04, 0.12)	16	−0.02	(−0.05, 0.01)
17	−0.01	(−0.07, 0.04)	18	−0.03	(−0.09, 0.02)	19	−0.00	(−0.03, 0.03)	20	0.04	(−0.02, 0.11)
21	−0.05	(−0.10, −0.01)	22	−0.02	(−0.06, 0.03)	23	0.02	(−0.01, 0.05)	24	−0.03	(−0.08, 0.03)
25	0.03	(−0.01, 0.06)	26	0.01	(−0.03, 0.04)	27	−0.02	(−0.09, 0.05)	28	−0.02	(−0.06, 0.02)
29	−0.05	(−0.10, 0.01)	30	0.00	(−0.03, 0.03)	31	0.01	(−0.03, 0.06)	32	−0.05	(−0.11, 0.02)
33	0.02	(−0.03, 0.07)	34	0.02	(−0.03, 0.07)	35	0.02	(0.00, 0.04)	36	−0.02	(−0.05, 0.01)
37	−0.01	(−0.03, 0.01)	38	0.01	(−0.05, 0.06)	39	0.01	(−0.04, 0.06)	40	−0.01	(−0.07, 0.04)
41	0.01	(−0.02, 0.03)	42	0.05	(−0.01, 0.11)	43	0.03	(−0.03, 0.09)	44	0.00	(−0.07, 0.07)
45	−0.02	(−0.06, 0.02)	46	−0.01	(−0.04, 0.03)	47	−0.01	(−0.07, 0.05)	48	0.00	(−0.02, 0.02)
49	−0.01	(−0.03, 0.01)	50	0.00	(−0.05, 0.05)	51	0.04	(−0.02, 0.10)	52	−0.01	(−0.07, 0.05)
53	−0.03	(−0.08, 0.02)	54	0.00	(−0.08, 0.08)	55	−0.01	(−0.03, 0.01)	56	0.03	(−0.05, 0.11)
57	0.00	(−0.04, 0.05)	58	−0.03	(−0.10, 0.04)	59	0.00	(−0.09, 0.09)	60	0.01	(−0.03, 0.05)
61	0.02	(−0.03, 0.07)	62	0.01	(−0.02, 0.03)	63	0.00	(−0.02, 0.02)	64	0.02	(−0.01, 0.05)
65	0.00	(−0.04, 0.04)	66	−0.04	(−0.11, 0.03)	67	0.03	(−0.04, 0.11)	68	0.04	(−0.01, 0.10)
69	0.02	(−0.02, 0.05)	70	0.01	(−0.01, 0.04)	71	−0.07	(−0.14, 0.01)	72	0.00	(−0.02, 0.02)
73	0.00	(−0.05, 0.05)	74	−0.02	(−0.08, 0.03)	75	−0.01	(−0.04, 0.02)	76	0.00	(−0.02, 0.03)
77	0.05	(−0.02, 0.12)	78	0.02	(−0.01, 0.06)	79	−0.01	(−0.09, 0.07)	80	0.03	(0.00, 0.07)
81	−0.01	(−0.03, 0.02)	82	0.01	(−0.04, 0.06)	83	0.03	(−0.03, 0.09)	84	0.03	(−0.05, 0.11)
85	0.00	(−0.06, 0.07)	86	0.01	(−0.01, 0.04)	87	0.00	(−0.05, 0.06)	88	0.00	(−0.04, 0.04)
89	0.00	(−0.05, 0.05)	90	0.02	(−0.03, 0.07)	91	−0.03	(−0.07, 0.01)	92	0.00	(−0.05, 0.06)
93	0.07	(0.00, 0.13)	94	−0.02	(−0.09, 0.04)	95	−0.02	(−0.07, 0.02)	96	−0.02	(−0.07, 0.02)
97	0.02	(0.00, 0.05)	98	−0.01	(−0.07, 0.06)	99	0.01	(−0.04, 0.06)	100	0.01	(−0.03, 0.05)
Known to be non-parallel											
101	0.05	(−0.01, 0.10)	102	0.07	(−0.01, 0.15)	103	0.07	(−0.01, 0.16)	104	0.10	(0.00, 0.19)
105	0.11	(0.06, 0.15)	106	0.02	(−0.04, 0.08)	107	0.11	(0.05, 0.17)	108	0.05	(0.00, 0.10)
109	0.08	(0.04, 0.13)	110	0.04	(−0.05, 0.13)						

TABLE B.13

Data for eight doses each tested in triplicate that demonstrate curvature in the dose-response relationship. These data are analysed in Section 7.3.2.1.

Dose	Response		
1.11	105.18	108.17	112.98
1.93	106.00	111.88	107.52
3.37	117.44	119.10	110.40
5.88	126.24	127.61	125.51
10.26	148.81	145.35	144.98
17.92	170.35	182.53	176.24
31.28	213.14	216.68	212.25
54.60	261.51	251.39	250.51

TABLE B.14

Data for ten doses each tested in triplicate from a 5PL dose-response relationship. These data are analysed in Section 7.3.2.2.

Dose	Response		
30000	3.93	3.98	3.94
12000	3.83	3.91	3.90
4800	3.84	3.75	3.77
1920	3.52	3.56	3.47
768	2.96	2.93	2.94
307.2	1.93	1.97	1.89
122.88	1.03	1.09	1.13
49.15	1.01	0.99	0.94
19.66	0.96	1.03	1.01
7.86	1.04	1.04	0.97

TABLE B.15

Data for demonstrating parallelism tests in Section 7.4.

Dose	Response					
	Reference standard			Test sample		
3	5133.79	5486.14	4377.44	2929.30	3206.04	2662.28
9	5234.71	5964.55	5391.95	3141.44	3326.26	3288.07
27	4962.20	5774.91	5222.17	3034.72	3523.68	4566.20
81	7443.38	8238.02	6957.12	5994.66	5289.99	5061.40
243	11604.84	12436.22	12110.39	10470.36	10691.20	10418.69
729	21421.30	19863.06	19460.92	18038.26	17941.20	18718.10
2187	26766.88	26441.36	25546.21	24545.70	24933.39	24888.21
6561	29359.61	29172.69	30080.83	27749.43	26824.46	27245.93
19683	30199.31	30064.39	28925.90	27761.56	27590.52	28383.53
59049	29408.74	29717.27	30093.11	27042.18	27827.25	28420.28

Bibliography

[1] Maurice S Bartlett. Properties of sufficiency and statistical tests. *Proceedings of the Royal Society of London. Series A-Mathematical and Physical Sciences*, 160(901):268–282, 1937.

[2] Douglas Bates, Martin Mächler, Ben Bolker, and Steve Walker. Fitting linear mixed-effects models using lme4. *Journal of Statistical Software*, 67(1):1–48, 2015.

[3] Bruno Boulanger, Walthère Dewé, Aurélie Gilbert, Bernadette Govaerts, and Myriam Maumy-Bertrand. Risk management for analytical methods based on the total error concept: conciliating the objectives of the pre-study and in-study validation phases. *Chemometrics and Intelligent Laboratory Systems*, 86(2):198–207, 2007.

[4] George EP Box and David R Cox. An analysis of transformations. *Journal of the Royal Statistical Society Series B: Statistical Methodology*, 26(2):211–243, 1964.

[5] RJ Briggs, R Nicholson, F Vazvaei, J Busch, M Mabuchi, KS Mahesh, M Brudny-Kloeppel, N Weng, PAR Galvinas, P Duchene, et al. Method transfer, partial validation, and cross validation: recommendations for best practices and harmonization from the global bioanalysis consortium harmonization team. *The AAPS Journal*, 16:1143–1148, 2014.

[6] David R Bristol. Probabilities and sample sizes for the two one-sided tests procedure. *Communications in Statistics-Theory and Methods*, 22(7):1953–1961, 1993.

[7] Richard K Burdick. Analytical procedure development and qualification, in Todd Coffey and Harry Yang (Eds.). *Statistics for biotechnology process development*, CRC Press, 2018.

[8] Richard K Burdick and Franklin A Graybill. *Confidence intervals on variance components*. CRC Press, 1992.

[9] Francis Bursa, Ann Yellowlees, Alka Bishop, Angela Beckett, Bassam Hallis, and Mary Matheson. Estimation of elisa results using a parallel curve analysis. *Journal of Immunological Methods*, 486:112836, 2020.

[10] Lai K Chan, Smiley W Cheng, and Frederick A Spiring. A new measure of process capability: Cpm. *Journal of Quality Technology*, 20(3):162–175, 1988.

[11] Todd Coffey. Design of experiments (DOE) for process development, in Todd Coffey and Harry Yang (Eds.). *Statistics for biotechnology process development*, CRC Press, 2018.

[12] David R Cox and David V Hinkley. *Theoretical statistics*. CRC Press, 1979.

[13] Jay L Devore. *Probability and statistics for engineering and the sciences, 9th edition*. Cengage Learning, 2016.

[14] Annette J Dobson and Adrian G Barnett. *An introduction to generalized linear models*. Chapman and Hall/CRC, 2018.

[15] Western Electric. *Statistical quality control handbook*. American Telephone and Telegraph Company, Chicago, IL, 1956.

[16] Timur V Elzhov, Katharine M Mullen, Andrej-Nikolai Spiess, and Ben Bolker. *minpack.lm: R Interface to the Levenberg-Marquardt Nonlinear Least-Squares Algorithm Found in MINPACK, Plus Support for Bounds*, 2023. R package version 1.2-4.

[17] Ph. Eur. Chapter 5.3: Statistical analysis of results of biological assays and tests. *European Pharmacopoeia: Version 11.0*, 2020.

[18] Edgar C Fieller. Some problems in interval estimation. *Journal of the Royal Statistical Society Series B: Statistical Methodology*, 16(2):175–185, 1954.

[19] David J Finney. *Statistical method in biological assay*. Number Ed. 3. Charles Griffin & Company, 1978.

[20] U.S. Food and Drug Administration. Analytical procedures and methods validation for drugs and biologics, 2015.

[21] U.S. Food and Drug Administration. 21 CFR § 601.2 applications for biologics licenses; procedures for filing, 2016.

[22] National Institute for Biological Standards and Control. Independent batch release testing at the NIBSC. `https://nibsc.org/control_testing/batch-release.aspx`. Accessed: 2024-05-30.

[23] Peter Goos and Bradley Jones. *Optimal design of experiments: a case study approach*. John Wiley & Sons, 2011.

[24] Paul G Gottschalk and John R Dunn. The five-parameter logistic: a characterization and comparison with the four-parameter logistic. *Analytical Biochemistry*, 343(1):54–65, 2005.

[25] Frank E Grubbs. Procedures for detecting outlying observations in samples. *Technometrics*, 11(1):1–21, 1969.

[26] IBM Corp. *IBM SPSS Statistics for Windows*, 2022.

[27] ICH. Guideline on validation of analytical procedures q2r2. *ICH Guidances*, 2024.

[28] Jiming Jiang and Thuan Nguyen. *Linear and generalized linear mixed models and their applications*, volume 1. Springer, 2007.

[29] John D Kalbfleisch and Ross L Prentice. *The statistical analysis of failure time data.* John Wiley & Sons, 2011.

[30] K Krishnamoorthy and Xiaodong Lian. Closed-form approximate tolerance intervals for some general linear models and comparison studies. *Journal of Statistical Computation and Simulation*, 82(4):547–563, 2012.

[31] David Lansky. Strategic bioassay design, development, analysis, and validation, in Todd Coffey and Harry Yang (Eds.). *Statistics for biotechnology process development*, CRC Press, 2018.

[32] Jerald F Lawless. *Statistical models and methods for lifetime data.* John Wiley & Sons, 2011.

[33] Stanley Levey and ER Jennings. The use of control charts in the clinical laboratory. *American Journal of Clinical Pathology*, 20(11_ts):1059–1066, 1950.

[34] Kung-Yee Liang and Scott L Zeger. Longitudinal data analysis using generalized linear models. *Biometrika*, 73(1):13–22, 1986.

[35] Ricardo A Maronna, R Douglas Martin, Victor J Yohai, and Matías Salibián-Barrera. *Robust statistics: theory and methods (with R).* John Wiley & Sons, 2019.

[36] Peter McCullagh. *Generalized linear models.* Routledge, 2019.

[37] Robert W Mee. Estimation of the percentage of a normal distribution lying outside a specified interval. *Communications in Statistics-Theory and Methods*, 17(5):1465–1479, 1988.

[38] Microsoft Corporation. *Microsoft Excel*, 2024.

[39] Minitab, LLC. *Minitab version 22.1*, 2024.

[40] Douglas C Montgomery. *Design and analysis of experiments.* John Wiley & Sons, 2017.

[41] Douglas C Montgomery. *Introduction to statistical quality control.* John Wiley & Sons, 2019.

[42] Douglas C Montgomery, Elizabeth A Peck, and G Geoffrey Vining. *Introduction to linear regression analysis.* John Wiley & Sons, 2021.

[43] David S Moore, George P McCabe, Layth C Alwan, and Bruce A Craig. *The practice of statistics for business and economics.* WH Freeman and Company, 2016.

[44] Lloyd S Nelson. The Shewhart control chart—tests for special causes. *Journal of Quality Technology*, 16(4):237–239, 1984.

[45] John Oakland. *Statistical process control.* Routledge, 2007.

[46] Gary W Oehlert. A note on the delta method. *The American Statistician*, 46(1):27–29, 1992.

[47] World Health Organization. Medicines: Good manufacturing practices. `https://www.who.int/news-room/questions-and-answers/item/medicines-good-manufacturing-processes`, 2015. Accessed: 2024-05-30.

[48] Quantics Biostatistics. *QuBAS Bioassay Software version 3.1.2*, 2024.

[49] R Core Team. *R: A Language and Environment for Statistical Computing*. R Foundation for Statistical Computing, Vienna, Austria, 2023.

[50] Sas Institute. *SAS Software version 9.4*, 2023.

[51] Franklin E Satterthwaite. An approximate distribution of estimates of variance components. *Biometrics Bulletin*, 2(6):110–114, 1946.

[52] Tara Scherder and Katherine Giacoletti. Continued process verification, in Todd Coffey and Harry Yang (Eds.). *Statistics for biotechnology process development*, CRC Press, 2018.

[53] Tim Schofield, Edwin van den Heuvel, Jane Weitzel, David Lansky, and Phil Borman. Distinguishing the analytical method from the analytical procedure to support the USP analytical procedure life cycle paradigm. *Pharmacopeial Forum*, 45, 11 2019.

[54] Andre Schuetzenmeister and Florian Dufey. *VCA: Variance Component Analysis*, 2024. R package version 1.5.0.

[55] JL Sebaugh and PD McCray. Defining the linear portion of a sigmoid-shaped curve: bend points. *Pharmaceutical Statistics: The Journal of Applied Statistics in the Pharmaceutical Industry*, 2(3):167–174, 2003.

[56] Samuel Sanford Shapiro and Martin B Wilk. An analysis of variance test for normality (complete samples). *Biometrika*, 52(3-4):591–611, 1965.

[57] Yueer Shi, Nicola Hulme, and Kieran O'Connor. Covalidation strategies to accelerate analytical method transfer for breakthrough therapies. *Pharmaceutical Technology*, 41(4):38–41, 2017.

[58] R Thorpe, M Wadhwa, and A Mire-Sluis. The use of bioassays for the characterisation and control of biological therapeutic products produced by biotechnology. *Developments in Biological Standardization*, 91:79–88, 1997.

[59] USP. Biological assay validation <1033>. *USP-NF*, 2013.

[60] USP. Design and development of biological assays <1032>. *USP-NF*, 2013.

[61] USP. Transfer of analystical procedures <1224>. *USP-NF*, 2013.

[62] USP. Statistical tools for procedure validation <1210>. *USP-NF*, 2018.

[63] USP. Analytical procedure life cycle <1220>. *USP-NF*, 2022.

[64] Jacobus Van Noordwijk. Bioassays in whole animals. *Journal of Pharmaceutical and Biomedical Analysis*, 7(2):139–145, 1989.

[65] Neil A Weiss and Carol A Weiss. *Introductory statistics.* Pearson London, 2017.

[66] James O Westgard, Patricia L Barry, Marian R Hunt, Torgny Groth, et al. A multi-rule Shewhart chart for quality control in clinical chemistry. *Clin Chem*, 27(3):493–501, 1981.

[67] Donald Wheeler and David Chambers. *Understanding statistical process control: third edition.* SPC Press, 2010.

[68] Hadley Wickham. *ggplot2: Elegant graphics for data analysis.* Springer-Verlag New York, 2016.

[69] Ann Yellowlees, Chris S. LeButt, Karie J. Hirst, Peter C. Fusco, and Kelly J. Fleetwood. Efficient Analysis of Dose-Time-Response Assays. *BioScience*, 63(6):490–498, 06 2013.

[70] Scott L Zeger, Kung-Yee Liang, and Paul S Albert. Models for longitudinal data: a generalized estimating equation approach. *Biometrics*, pages 1049–1060, 1988.

[71] Kirkwood, Thomas BL. Geometric means and measures of dispersion. *Biometrics*, 908–909, 1979.

Index